The R.A.M.S. Library of Alchemy

Volume 5

Three Works of George Ripley:

The Compound of Alchemy

The Marrow of Alchemy

Liber Secretissimus

R.A.M.S. Publishing Company

Three Works of George Ripley:

The Compound of Alchemy

The Marrow of Alchemy

Liber Secretissimus

Produced by

Restorers of Alchemical Manuscripts Society

R.A.M.S. Publishing Company

R.A.M.S. Publishing Company
117 Rutherford Lane
Stuarts Draft VA 24477

Copyright © 2015 R.A.M.S. Publishing Company

All rights reserved. No part of this publication may be reproduced or transmitted in any form or by any means, electronic or mechanical, including but not limited to any information storage and retrieval system, without written permission from R.A.M.S Publishing Company. Reviewers may quote brief passages.

First Edition 2015

ISBN-13 **978-1508614999**
ISBN-10 **1508614997**

Image Processing by Philip N. Wheeler

This book is sold for informational purposes only. Neither the publisher nor the editor shall be held accountable for the use or misuse of the information in this book.

Printed in the United States of America

A Portion of The Ripley Scroll, 16th Century

Table of Contents

Introduction...7
THE COMPOUND OF ALCHEMY.............................11
The Marrow of Alchemy............................124
LIBER SECRETISSIMUS..............................215
A Word from the Publisher 237

Introduction

Philip N. Wheeler

This volume contains the three works from the R.A.M.S. Library that are attributed to George Ripley:

The Compound of Alchemy

The Marrow of Alchemy

Liber Secretissimus

Sir George Ripley (circa 1415 – 1490) was an English Alchemist, author and Augustine canon. His Alchemical writings were studied by many notable people, including Robert Boyle (considered to be the first modern chemist), John Dee, and Isaac Newton.

The Compound of Alchemy; or, the Twelve Gates leading to the Discovery of the Philosopher's Stone (Liber Duodecim Portarum) was published in 1591 (London: Thomas Orwin). It was one of Ripley's most popular works.

The Marrow of Alchemy, or *Medulla philosophiæ chemicæ*, was published in 1614 (Francofurti: J. Bringer).

Liber Secretissimus has the subtitle, "The Whole Work of the Composition of the Philosophical Stone and Grand Elixir, and of the First Solution of the Grosse Bodies."

More than 200 manuscripts are attributed to Ripley. Most of them have never been published.

Dedicated to Hans W. Nintzel,
American Alchemist
and
Founder of the
Restorers of Alchemical Manuscripts Society
(R.A.M.S.)

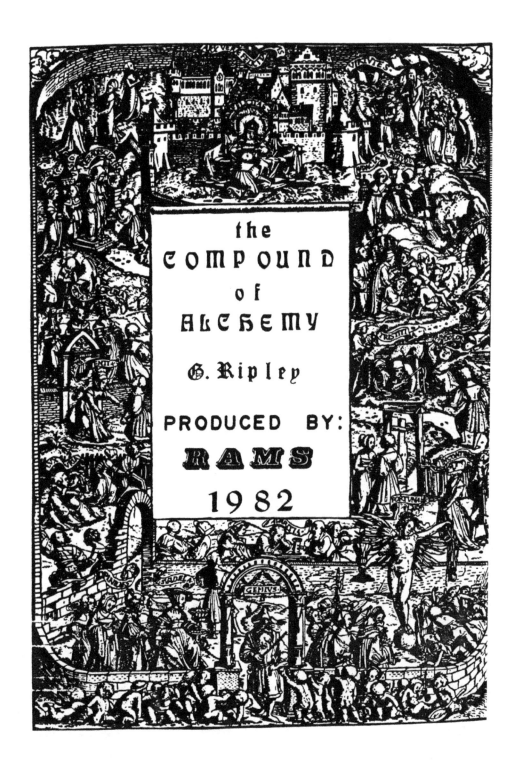

THE COMPOUND OF ALCHEMY

OR

The Ancient Hidden Art of Alchemie; Containing the right and perfect means To make the Philosophers Stone

Aurum Potabile, with other

Excellent Experiments.

Divided into Twelve Gates

First written by the learned and rare Philosopher of our Nation GEORGE RIPLEY, sometimes Chanon of Bridlington in Yorkeshire and Dedicated to K. Edward the 4th. whereunto is adjoined his Epistle to the King his Vision, his Wheel, and other his works never before published:

Set forth by Raph Rabbards, gentleman, Studious and expert in Alchemical Arts

Pulchrum pro Patria pati.

Imprinted by Thomas Orwin.

LONDON

1591.

THE COMPOUND OF ALCHEMY

To the Right Honourable, Worshipful, and Worthy Gentlemen of England, and other Learned and Industrious Students in the Secrets of Philosophy.

Having reserved the copy hereof (Right Honourable and gentle Readers) these forty years for many secret uses, corrected by the most learned of our time, and feeling myself, now through age declining, and otherwise hindered with troublesome suites in law, which constrained me to discontinue these and such other commendable practises, for the benefit and defence of my Prince and Country, I have thought good to publish the same, the rather for that there are but a few copies left, and those for the most part corrupted by negligence, or mistaking of ignorant writers thereof; being partly incouraged hereunto by the learned Philosopher SENECA, who accompleth it an Act Meritorious to preserve or revive Science from the Cinders, and to eternize vertuous acts with perpetual memory.

Finding it stranger that so excellent a Monument as this most rare and learned work of GEORGE RIPLEY, should so long lie hidden in obscurity, and pass from hand to hand a hundred and fifty years without utter defacing; seeing that many notable works published have, in far shorter time perished.

He lived in the time of King Edward the 4th., and Richard the Usurper, in great fame and estimatiom, for his rare knowledge in these secrets: And to utter his further commendation in a word, if this his work be pursued with the eye of judgement, and practised by those that are experienced, I shall not need to fear the loss of my labour, when some of my countrymen which cannot attain into the highest hidden mysteries shall yet find many things both pleasant and profitable, divers ways serviceable to kingdoms, states, and common weales; And if one among ten thousand can hit the mark, (whereas the whole world hath aimed) we shall not hereafter need to seek out the Sands of the Ganges, for that which lieth hid in the secrets of Ripley; which I offer here to the view of the learned, and have presumed the publishing hereof, chiefly for the benefit of my countrymen; and especially to satisfy the often and importunate request of many my learned good friends, not doubting but that the skillful will yield me my due, howsoever the ignorant shall esteem thereof; which if I shall thereby be further encouraged to impart some other rare experiments of Distillations and Fire-works of great service, not hitherto committed in writing or put in practice by any of our nation, although of late some mear toys have been highly admired, and extraordinarily rewarded; The charge whereof will be found utterly lost, when perfect trial shall be made of their slender use and force. To conclude, if this my good intent shall be answerable to your expectations, I have obtained the fulness of my desires.

Yours in the futherance of Science,

                                              Raph Rabbards.

A Preface

by George Ripley, Chanon of Bridlington, Yorkeshire.

    This small volume is presented to the sincere student in hopes that it will inspire his Studies and lead him to the true goal that he seeks.
    With the blessing of Almighty God, all students who undertake the Great Work must beware of naysayers, unworthy consorts, leaguers, and adherents, or other their Ministers, spreaders and maintainers of lies; under the shelter and covert of which flourishing Palme, all true christians have been, and (I hope) long shall be (by the Continuance of your Majesties most bountiful and gracious especial favour) protected and shrowded, from the burning heat of the sharp persecutions of all malicous Enemies thereof the which God of his great mercie Grant.
Pondering I say (most high and mighty Princess) the manifold imminent dangers, your sacred Majesty by the Omnipotence wonderful providence of Almighty God, with more than human Virtue, and feminine patience sustained; and in the end (contrarie to all expectation) escapeing the rage, furie, Tryannical attempts, and secret devilish practices of your Highnesses mortal Enemies (utter professed unreconcileable adversaries) to the eternal truth; Whereof myself was

an eye witness, and so far privie of some of the most mischievous intended conspiracies, as for my faithfull endeavours of such rare secret services as were by me effected to prevent the same; I tasted (amongst the rest of such as then honoured, reverenced, and by bounded duty loved your Majesties rare Virtues) So great extremitie of imprisonment, and other hard usage many ways, as escaping with life, (by timely and happy alteration of the State) I felt long after the pains of those torments, whereby my health in 20 years after was extremely impared: But when I consider on the other side withall, the strange and miraculous Heroical Actions both martial and politick which have been (since in the time past of your Highness most prosperous reign) beyond all expectations performed with singular government and incomparable magnamimite, far surpassing all human Wisdom and mans force (whereof posteritie shall want no due records of worthy Registers) I cannot but forget all sorrow, and exceedingly rejoice, assuring and perswading myself God hath ordained your Majestie to accomplish yet far greater matters for his Glory and your own eternal memory, than time hath yet revealed. For the better accomplishment whereof, every dutiful subject and good Patriots ought in time of tranquilitie prepare and furnish themselves according to their several gifts, abilities, and different faculties; to further and set forth a work so great as is likely to fall out by all consequence by reason of this your Majesties most happy reign: Viz, the Nobilitie with learning, Judgement, and experience, for council and advice, as well for warlike discipline as for evil Government: For that the one cannot long continue without the other and Gentlemen with ingeneous devices and stratagies of war both for sea, and land, and the whole soldierie of all sorts such activities, and the practise of such sorts of weapons as they shall addict themselves unto, or be found fit to serve withall, whereby every man of what degree or quality soever professing the wars, may in time of cessation of Arms, some in one sort, some in another, enable themselves for the defence of their Country, against they shall be employed; for without appointment and calling, no man ought to intrude or offer himself, in such sort as is now too commonly used; which is the cause that so many andatious insufficient blind bastards are set a work: Whilst the modest, skillfull, and experienced live retired, attending their vocation.

Note. Here this Preface ends; whether it is complete or not, I have no way of knowing. I presume it is by George Ripley, but again this is not certain, at best it is not important, and pertains not to the Twelve Gates. D. H.

The Vision of Sir George Ripley.

Chanon of Bridlington.

When busie at my book upon a certain night,
This Vision here exprest appeared unto my dimmed sight,
A Toad full rudde I saw did drink the juice of grapes so fast
Till overcharged with broath, his bowels all to brast,
And after that from poisened bulke, he cast his venome fell,
For grief and pain whereof his members all began to swell,
With drops of poisoned sweat approaching thus his secret den
And from which in space a golden humour did ensue,
Whose falling drops from high did stain the soil with ruddy hue
And when his corps, the force of vital breath began to lack,
This dying toad became forth with like coal for colour black;
Thus drowned in his proper veins of poisoned flood,
For term of eightie days and four he rotting stood:
By tryal then this venome to expell I did desire,
For which I did commit his carkase to a gentle fire;
Which done, a wonder to the sight, but more to be rehearst,
The Toad with colours rare through every side was pearced;
And white appeared when all the sundry hues were past
Which after being tincted, rudde for evermore did last;
When of the venome handled thus a medicine I did make,
Which venom kills, and saveth such as venome chance to take,
Glory be to him the granter of such secret ways
Dominion, and Honour, both with worship and with praise.

AMEN.

The Preface.

O   High, incomprehensible and glorious majesty,
Whose luminous beams obtaineth our speculation

O   Trinehood in persons, O   Onehood in diety,
Of Hierarchical Jubilists gratutant glorification

O   Petious Purifier of Souls and pure perpetuation,
O   deniant from danger, O drawer most debonaire,
From this troublesome vale of vanity, O our Exalter.

O Power, O Wisdom, O Goodness inexplicable,
Support me, teach me, and be my Governor,
That never my living to thee be displicable,
But that I acquit me to thee as true professor.
At this beginning Good Lord hear my prayer.
Be nigh with grace for to inforce my will,
Grant wit that I may mine intent fulfill.

Most curious cofer and copious of all treasures,
Thou art he from whom all goodness doth descend,
To man and also to every creature
Thine handy-work therefore vouchsafe to defend,

That we no time in living here misspend,
With troth here grant us our living so to win,
That into no danger of sinfullness we rin.

And forasmuch as we have for thy sake
Renounced our wills, the world and fleshly lust,
As thine own professors us to thee take,

Sith in thee only dependeth all our trust,
We can no further, to thee incline we must;

Thy secret treasure vouchsafe unto us,
Show us thy secrets and to us be bounteous.

And amongst others which be profest to thee
I me present as one with humble submission,
Thy servant beseeching that I may be
And true in living according to my profession,
In order Chanon Regular of Bridlington
Beseeching thee Lord thou wilt me spare
To thy true servants thy secrets to declare.

In the beginning when thou mad'st all of naught,
A globous matter and dark under confusion,
By him the beginning marvelously was wrought,
Containing naturally all things without division;
Of which in six days he made clear distinction
As Genesis apertly doth record
Then Heaven and Earth were perfect by his word.

So through his will and power, out of one mass,
Confused; was made each thing, that being is,
But afore in Glory as maker he was,
Now is and shall be without end I wis,
And purified souls up to his bliss
Shall come a principle this may be one,
For declaring of our precious Stone.

For as of one mass was made all thing,
Right so in our practice must it be,
All our secrets of one Image must spring,
As in Philosophers books who list to see,
Our stone is called the lesser world, one and three,
Magnesia also of sulphur and mercurie,
Proportionate by nature most perfectly.

But many one marveleth and marvel may,
And museth such a marvelous thing,
What is our stone; sith philosophers say
To such as ever be it in seeking
Fouls and fishes to us doth it bring
Every man it hath, and it is in every place,
In thee in me, and in each thing, time and space.

To this I answer that mercury it is I wis
But not the common called Quicksilver by name,
But Mercury without which nothing being is,
All philosophers record and truly same the same,
But simple searchers putteth them in blame,
Saying they hid it but they be blame worthy,
Which be no clearks and meddle with philosophy.

But though it mercury be, yet wisely understand,
Where in it is, and where thou shalt it seech,
Else I counsel thee take not this work in hand,
But listen to me, for truly I will thee teach,

Which is this mercury most profitable,
Being to thee nothing deceiveable.

It is more near in some things than in some;
Therefore take heed what I to thee write,
For if knowledge to thee never come,
Therefore yet shalt thou me not twite,
For I will truly now thee excite
To understand well Mercuries three,
The keys which of this science be.

Raymond his menstrues doth them call,
Without which truly no truth is done;
But two of them be superficial,
The third essential of Sun and Moon,
Their properties I will declare right soon,
And Mercury of metals essential,
Is the principal of our stone material.

In Sun and Moon our menstrue is not seen,
It appeareth not by effect to sight
That is the stone of which we mean,
Who so our writings convieveth aright,
It is a Soul, a substance bright,
Of Sun and Moon a subtile influence,
Whereby the earth revieveth resplendence.

For what is Sun and Moon saith Avicin,
But earth which is pure white and red;

Take from it the said clearness, and then
That earth will stand but a little stead;
The whole compound is called our Lead;
The Quality of clearness from Sun and Moon doth come,
These are our menstrues both all and some.

Bodys with the first we calcine naturally
Perfect, but none which been unclean
Except one which is usually
Named by philosophers the Lyon Green;
He is the meane Sun and Moon between
Of winning tincture with perfectness
As Geber thereunto beareth witness.

With the second which is an humidity
Vegetable, reviving that earst was dead,
Both principles materials must loosed be
And formalls, else stand they little stead;
These menstrues therefore know I thee read
Without the which neither true calcination
Done may be, nor true dissolution.

With the third humiditie most permanent
Incombustible and unctious in his nature,
Hermes Tree into ashes is burnt,
It is our natural fire most sure,
Our mercury, our sulphur, our tincture, pure,
Our soul, our stone born up with the wind
In earth engendered, bear this in mind.

This stone also tell thee I dare,
Is the vapour of metals potential,
How thou shalt get it, thou must beware,
For invisible truly is this menstrual
By separation of elements it may appear
How be it with the second water philosophical
To Sight, in form of water clear.

Of this menstru by labour exuberate
With it may be made sulphur of nature,
If it be well and kindly acuate
And circulate into a spirit pure,
Then to dissolve thou must be sure,
Thy base with it in divers wise
As thou shalt know by thy practise.

That point therefore in his due place
I will declare with other moe,
If God will grant me grace and space,
And me preserve in life from woe,
As I thee teach look thou do 50;
And for first ground principal,
Understand thy waters menstrual.

And when thou hast made true calcination,
Encreasing not wasting moisture radical,
Until thy base by oft subtilation,
Will lightly flow as wax upon metal

Then loose it with thy vegetable menstrual,
Till thou have oil thereof in colour bright
Then is that menstru visible to sight.

And oil is drawn out in colour of gold,
Or like thereto out of our fine red lead,
Which Raymond said when he was old,
Much more than gold would stand him in stead:
For when he was for age nigh dead,
He made thereof Aurum Potabile,
Which him revived as men might see.

For so together may they be circulate,
That is to say oil and the vegetable menstrual,
So that it be by labour exuberate,
And made by craft a stone celestial,
Of nature so fiery that we it call
Our Bazeliske or our Cockatrice,
Our Great Elixir most of price.

Which as the sight of Baseliske his object
Killeth, so slayeth it crude Mercury,
When thereupon he is project,
In twinckle of an eye most suddenly
That Mercury then teineth permanently
All bodies to Sun and Moon perfect.
Thus guide thy base both red and white.

Aurum Potabile thus is made,

Of gold not commonly calcinate,
But of our tincture that will not fade,
Out of our base drawn with the menstrue circulate
But natural calcination must algate
Be made, ere thy gold dissolved may be,
That principle first therefore I will teach thee.

But into chapters this treatise I will divide,
In number twelve with due recapitulation,
Superfluous rehearsals I'll lay aside,
Intending only to give true information;
That by my writing whoso guided will be
Of his intent perfectly speed shall he.

The first chapter shall be of natural calcination,
The second of dissolution secret and philosophical
The third of our elemental seperation,
The fourth of conjunction matrimonial,
The fifth of putrifaction follow shall,
Of congealation albificate shall be the sixth,
Then of cibation the seventh shall follow next.

The secret of sublimation the eight shall show,
The Nineth shall be of fermentation;
The tenth of our exaltation I trow;
The eleventh of our marvellous multiplication;
The twelveth of projection, then recapitulation;
And so this treatise shall take end.
By the help of God as I intend.

Finis.

Of Calcination
The First Gate.

Calcination is the Purgation of our stone,
Restoring also of his natural heat,
Of radical humidity it looseth none,
Inducing solution into our stone most meet
After Philosophie I you behight
Do, but not after the common guise
With Suiphures and salts prepare in divers wise.

Neither with corrosives nor with fire alone,
Neither with vinegar nor with water ardent,
Nor with vapour of lead our stone
Is calcined according to our intent:
All those to calcining which be so bent,
From this hard science withdraw their hand,
Till they OUR calcining better understand.

For by such calcination their bodies be spent,
Which minisheth the moisture of our stone,
Therefore when bodies to powder be brent,
Dry as ashes of tree or bone,
Of such calxes then will we none,
For moisture we multiplie radicall
Incalcining minishing none at all.

And for a sure ground of our true calcination,
Work wittely only kind with kind;

For kind unto kind hath appetitive inclination,
Who knoweth not this in knowledge is blind,
He may forth wander as mist in the wind,
Wotting never with profit where to light,
Because he conceives not our words aright.

Join kind to kind therefore as reason is,
For every burgeon answers his own seed,
Man getteth man, a beast a beast I wis,
Further to treat of this it is no need;
But understand this point of thou wilt speed,
Each thing is first calcined in his own kind,
This well conceived fruit shalt thou find.

And we make caix unctuous both white and red,
Of three degrees or our base be perfect.
Fluxible as wax, else stand they no sted.
By right long process as Philosophers do write,
A year we take or more to our respite;
For in less space our calxes will not be made,
Able to teyne with colour that will not fade.

And for thy proportion thou must beware,
For therein maist thou be beguiled,
Therefore thy work that thou not mare,
Let thy bodie be subtilly filled
With mercury as much then so subtilled
One of the Sun, two of the Moon
Till altogether like pap be done.

Then make mercury four to the Sun,
Two to the moon as it should be,
And thus thy work must be begun,
In figure of the trinity,
Three of the bodie and of the spirit three,
And for the unity of the substance spiritual
One more than of the substance corporal.

By Raymonds reportory this is true.
Proportion there who list to look,
The same my Doctor to me did show
But three of the spirit Bacon took,
To one of the bodie for which I awoke,
Many a night ere I it wist,
And both be true take which you list.

If the water also be equal in proportion
To the earth, with heat in due measure,
Of them shall spring a new burgeon,
Both white and red in pure tincture,
Which in the fire shall ever endure,
Kill thou the Quick the dead revive,
Make trinity unite without any strife.

This is the surest and best proportion,
For there is least of the part spiritual
The better therefore shall be solution,
Than if thou did with water swall,

Thine earth over glutting which loseth all,
Take heed therefore to potters loam,
And make thou never to neshe thy womb.

The loam behold how it tempered is,
The mean also how it is calcined,
And ever in mind look thou bear this;
That never thine earth with water be suffocate,
Dry up thy moisture with heat most temperate,
Help dissolution with moisture of the Moon,
And Congealation with the Sun, then hast thou done.

Four natures into the fifth so shalt thou turn,
Which is a Nature most perfect and temperate,
But hard it is with thy bare foot to spurn
Against a bar of iron, or steel new acuate,
For many do so which be infatuate,
When they such high things take in hand,
Which they in no wise do understand.

In eggs, in vitriol, or in blood,
What riches wend they there to find,
If they Philosophy understood
They would not in working be so blind,
Gold and Silver to seek out of kind.
For like as fire of burning the principle is,
So is the principle of gilding, gold I wis.

If thou intend therefore to make,

Gold, and Silver by craft of our philosophie
Thereto neither eggs nor blood thou take
But Gold and Silver which naturally
balcined wisely and not manually,
A new generation will forth bring,
Encreasing their kind as doth everything.

And if it true were that profit might be,
In things which be not metalline,
In which be colours pleasant to see,
As in blood, eggs, hair, urine, or wine,
Or in mean minerals digged out of the mine,
Yet must their elements be putrified and separate
And with elements of perfect bodies be dispousate.

But first of these elements make thou rotation,
And into water thine earth turn first of all,
Then of thy water make air by levigation
And air make fire, then master will I call thee
Of all our secrets great and small:
The wheel of elements then canst thou turn about,
Truly conceiving our writings without doubt.

This done, go backwards turning the wheel again,
And into thy water turn thy fire anon,
Air into earth, else laboureth thou in vain
For so to tempernient is brought our stone,
And natures contrari-wise four are made one,
After they have three times been circulate,

And also thy base perfectly consumate.

Thus under thy moisture of the Moon,
And under the temperate heat of the Sun,
Thine elements shall incenerate soon.
And then thou hast the maistrie won;
Thank God thy work was then so begun,

For there thou hast one token true
Which first in blackness to thee will show.

The head of the crow that token call we,
And some men call it the crows bill,
Some call it the ashes of Hermes tree,
And thus they name it after their will;
Our toad of the earth which eateth his fill,
Some name it by which it is mortificate,
The spirit of the earth with venom intoxicate.

But it hath names I say to thee infinite
For after each thing that black is to sight,
Named it is till it waxeth white.
Then hath it names of more delight,
After all things that been full white,
And the red likewise after the same,
Of all things red doth take the name.

At the first gate now art thou in,
Of our philosophers castle where they dwell,

Proceed wisely that thou may win,
In at more gates of that castle
Which castle is round as any bell,
And gates it hath eleven yet more,
One is conquered, now to the second go.

            End of the First Gate.

Of Dissolution
The Second Gate

Of Dissolution now will I speak a word or two
Which sheweth out what erst was hid from sight,
And maketh intenuate things that were which also,
By virtue of our first menstrue clear and bright,
In which our bodies eclipsed been of light,
And of their hard and dry compaction subtilate,
Into their own first matter kindly retrogradate.

One in gender they be, and in number two,
Whose Father is the Sun, the Moon the Mother,
The Mover is Mercury, these and no more
Be our magnesia, our Adropp, and none other
Things there be, but only sister and brother,
That is to mean agent and patient,
Sulphur and Mercury coessential to our intent.

Betwixt these two equalitic contrarious
Ingendred is a mean most marvelously
Which is our Mercury and menstrue unctuous
Our secret sulphur working invisibly.
More fiercely than fire burning the bodie,
Dissolving the bodie into water mineral,
Which night for darkness in the North we do call.

But yet I trow thou understandest not utterly
The very secret of the Philosophers Dissolution,

Therefore conceive me I counsel thee wittely,
For the truth I will tell thee without delusion;
Our solution is cause of our congelation, For dissolution on the one side corporal Causeth congelation on the other side spiritual.

And we dissolve into water which wetteth no hand,
For when the earth is integrately incinerate,
Then is the water congealed; this understand
For the elements be so together concatenate,
That when the bodie is from his first form alterate
A new form is induced immediately,
For nothing being without all form is utterly.

And here a secret to thee I will disclose,
Which is the ground unto our secrets all,
And it not known thou shalt but loose,
Thy labour and cost both great and small,
Take heed therefore in error that thou not fall
The more thine earth, and the less thy water be,
The rather and better solution shalt thou see.

Behold how ice to water doth relent,
And so it must for water it was before,
Right so again to water our earth is went
And water thereby congealed for evermore,
For after all philosophers that ever were bore,
Each metal was once water mineral,
Therefore with water they turn to water all.

In which water of kind occasionate,
Of Qualities been repugnant and diversitie,
Things into things must therefore be rotate,
Until diversitie be brought to perfect unitie;
For scripture recordeth when the earth shall be
Troubled, and into the deep sea shall be cast
Mountains and bodies likewise at the last.

Our bodies be likened conveniently
To mountains, which after high planets we name,
Into the deepness therefore of Mercury
Turn them, and keep thee out of blame,
For then shalt thou see a noble game,
How all shall become powder as soft as silk,
So doth our rennet kindly kurd up our milk.

There hath the bodies their first form lost,
And others been induced immediately
Then hast thou well bestowed thy cost;
Whereas others uncunningly must go by,
Not knowing the secrets of our philosophy.
Yet one point more I must tell thee,
How each bodie hath dimensions three.

Altitude, Latitude, and also Profunditie,
By which all gates turn we must our wheel,
Knowing that thine entrance in the West shall be,
Thy passages forth to the North if thou do well,

And there thy lights lose their lights each deele;
For there thou must abide by ninetie nights
In darkness of purgatorie withouten lights.

Then take thy course up to the East anon,
By colours passing variable in manifold wise
And then be winter and vere nigh overgone,
To the East therefore thine ascending devise,
For there the Sun with daylight doth uprise,
In Summer, and there disport thee with delight,
For there thy work shall become perfect white.

Forth from the East into the South ascend,
And set thee down there in the chair of fire,
For there is harvest; that is to say an end
Of all this work after thine own desire
There shineth the Sun up in his hemisphere,
After the eclipses, in redness with glory
As king to reign upon all metals and mercury.

And in one glass must be done all this thing
Like to an egg in shape and closed well,
Then must thou know the measure of fireing,
The which unknown thy work is lost each deele:
Let never thy glass be hotter than thou maist feel,
And suffer still in thy bare hand to hold,
For fear of loosing, as philosophers have told.

Yet my doctrine further attend

Beware thy glass thou never open nor move
From the beginning till thou have made an end;
If thou do contrarie thy work may never cheve
(acheave)
Thus in this chapter which is but brief,
I have taught thee thy true solution;
Now to the third gate go, for this is won.

        End of the Second Gate.

Of Separation

The Third Gate

Separation doth each part from other divide,
The subtile from the gross, the thick from the thin.
But separation manual look thou set aside,
For that pertains to fools that little good doth win,
But in our separation Nature doth not blind
Making division of Qualities elemental,
Into fifth degree till they be turned all.

Earth is turned into water under black and blue,
And water after into air under very white,
Then air into fire, elements there be no more,
Of these is made our stone of great delight.
But of this Separation is called by Philosophers definition
Of the said four elements tetrative dispersion.

Of this separation I find a like figure,
Thus spoken, by the prophet in the Psalmodie,
God brought out of a stone a flood of water pure,
And out of the hardest rock oil abundantly,
So out of our stone precious if thou be witty,
Oil incombustable, and water thou shalt draw,
And there abouts at the coal thou needst not to blow.

Do this with heat easie and nourishing,
First with moist fire and after with dry,
The flegm with patience outdrawing,
And after that the other natures wittely.
Dry up thine earth until it be thirsty,
By calcination else labourest thou in vain,
And then make it dry up the moisture again.

Separation thus must thou oftentimes make,
Thy waters dividing into parts two,
So that the subtile from the gross thou take
Till the earth remain beneath in colours blue
That earth is fixed to abide all woe,
The other part is spiritual and flying,
But thou must turn them all into one thing.

Then oil and water with water shalt distill
And through her help receive moving
Keep well these two that thou not spill
Thy work for lack of due closing
And make thy stopple of glass, melting
The top of thy vessel together with it,
Then Philosopher-like it is up shut.

The water wherewith thou mayst revive the stone,
Look though distill before thou work with it,
Often times by itself alone,
And by this sight thou shalt wit,
From feculent feces when it is quit;

For some men can with saturn it multiplie
And other substance which we defie.

Distill it therefore till it be clean,
And thin like water as it should be,
As heaven in colour bright and sheen,
Keeping both figure and ponderositee
Therewith did Hermes moisten his tree;
Within his glass he made it grow upright
With flowers discoloured beautiful to sight.

This water is like to the venomous tire,
Wherewith the mighty triacle is wrought
For it is Poison most strong of ire,
A stronger poison cannot be thought,
At Pothecaries often therefore it is sought,
But no man by it shall be intoxicate,
From the time it is into medicine elixerate.

For then as is the Triacle True,
It is of poison most expulsive,
And in his working doth marvels shew
Preserving many from death to life,
But that thou meddle it with no corrosive,
But choose it pure and Quick running
If thou thereby wilt have winning.

It is a marvellous thing in kind,
And without it nothing can be done,

Therefore Hermes called it his wind,
For it is upflying from Sun and Moon,
And maketh our stone fly with it soon,
Reviving the dead and giving life,
To Sun and Moon, husband and wife.

Which if they were not by craft made quick,
And their fatness with water drawn out,
And so the thin diservered from the thick,
Thou shouldst never bring this work about,
If thou wilt therefore speed without doubt
Raise up the birds out of their nest
And after again bring them to rest.

Water with water accord will, and ascend,
And Spirit with spirit, for they be of one kind,
Which after they be exalted make to descend
So shalt thou divide that, which Nature erst did bind,
Mercury essential turning into wind,
Without which natural and subtile separation,
May never be complete profitable generation.

Now to help thee at this gate,
The last secret I will declare to thee,
Thy water must be seven times sublimate,
Else shall no kindly Dissolution be,
Nor putrifying shalt thou none see;
Like liquid pitch, nor colours appearing
For lack of fire within the glass working.

Four fires there be which thou must understand
Natural, unnatural, against Nature also,
And elemental which doth burn the brand;
These four fires use we no more,
Fire against nature must do thy bodie woe,
This is our Dragon as I thee tell,
Fiercely burning as the fire of Hell.

Fire of nature is the third menstrual,
That fire is natural in each thing;
But fire occasionate we call unnatural,
As heat of ashes, and blanes for putrifying,
Without these fires thou maist nought bring
To Putrifaction, for to be separate
Thy matters together proportionate.

Therefore make fire thy glass within,
Which burneth the bodie much more than fire
Elemental, if thou wilt win
Our secrets according to thy desire;
Then shall thy seeds both rot and spire
By help of fire occasionate,
That kindly after may be separate.

Of Separation the Gate must thus be won,
Towards the Gate of secret conjunction,
That furthermore yet thou maist proceed
Into the Castle which will thee innerlead.

Do after my councel if thou wilt speed,
With two strong locks this Gate is shut,
As consequently thou shalt well wit.

        End of the Third Gate.

The Fourth Gate.

Of Conjunction.

After the chapter of natural Separation,
By which the elements of our stone disevered, be,
The chapter here followeth of secret conjunction,
Which natures repugnant joineth to perfect unitie,
And so them knitteth that none from others may fly
When they be fire shall be examinate,
They be together so surely conjungate.

And therefore Philosophers gave this difinition,
Saying this conjunction is nothing else
But of principles a co-equation as others tells;
But some men with mercury that Pothecaries sells
Medleth bodies, which cannot divide,
Their matter, and therefore they slip aside.

For until the time the soul be separate
And cleansed from his original sin
With the water, and throughly spiritualizate
The true conjunction maist thou never begin;
Therefore the soul first from the bodie twyne
Then of the corporall part and of the spiritual
The soul shall cause conjunction perpetual.

Of two Conjunctions Philosophers mention make,

Grosse when the body with mercury is reincrudate,
But let this pass, and so the second heed take.
Which as I said is after Seperation celebrate
In which the parties be left with least to colligate,
And so promoted into most perfect temperance, That
never after amongst them may be repugnance.

Thus causeth Separation true conjunction to be had,
Of water and air, with earth and fire,
But that each element into other may be led,
And so abide forever to thy desire,
Do as do dawbers with clay or mire,
Temper them thick and make them not to thin,
So do updrying, thou shalt the rather win.

But manners there be of our conjunction three,
The first is called by Philosophers diptative,
The which betwixt the agent and patient must be,
Male and female, mercury, and sulphur vive
Matter and form, thin and thick to thrive,
This lesson will help thee without any doubt,
And our conjunction truly to bring about.

The second manner is called Triptative,
Which is conjunction, made of things three,
Of body, soul, and spirit, that they may not strive,
Which trinitie thou must bring to unite,
For as the soul to the spirit the bond must be,
Right so the bodie the soul to him must knit,

Out of thy mind let not this lesson flit.
The third mariner and also the last of all,
Four elements together which join to abide,
Tetraptative certainly Philosophers do it call,
And specially Guido de Montanio whose same goeth wide,
And therefore in most laudable manner this tide,
In our conjunction four elements must aggregate
In due proportion, which first asunder were separate.

Therefore like as the woman hath veins fifteen,
And the man but five to the act of their secunditie,
Reauired in our conjunction first I mean,
So must the man our Sun have of his water three
And nine his wife, which three to him must be;
Then like with like will joy have for to dwell,
More of conjunction we needeth not to tell.

This chapter I will conclude right soon therefore,
Gross conjunction charging thee to make but one,
For seldom have strumpets children of them bore,
And so thou shalt never come out by our stone,
Without thou let the woman lie alone,
That after the once have conveived of the man,
Her matrix be shut up from all other than.

For such as add ever more crude to crude
Opening their vessel letting their matters keele
The sperme conceived they nourish not but delude
Themselves, and spill their work each deele,

If thou therefore have list to do well
Close up thy matrix and nourish the seed
With heat continual and temperate if thou wilt speed.

And when thy vessel hath stood by months five,
And clouds and eclipses be passed each one,
The light appearing, increase thy heat then below,
Until maist thou open thy glass arione,
And seed thy child which is now bore,
With milk and meat aye more and more.

For now both moist and dry is so contemperate,
That of the water earth hath received impression
Which never (after that) asunder may be seperate
And right so water to earth hath given ingression
That both together to dwell have made profession
And water of earth hath purchased a retentive,
They four make one never more to strive.

Thus in two things all our intent doth hing,
In dry and moist, which be contraries two,
In dry, that it the moist to fixing bring,
In moist, that it give liquefaction to the earth also,
Then of them thus a temperment may forth go
A temperment not so thick as the body is,
Neither so thin as water withouren miss.

Loosing and knitting thereof be principles two
Of this hard science, and poles most principal,

Howbeit that other principles be many more,
As shining fanes, which show I shall;
Proceed therefore unto another wall
Of this strong castle of our wisdom,
That in at the fourth Gate thou maist come.

        End of the Fourth Gate

Of Putrefaction.

The Fifth Gate.

Now we begin the chapter of Putrefaction,
Without which pole no seed may multiply;
Which must be done only by continual action
Of heat in the bodie, moist not manually;
For bodies else may not be altered naturally,
Sith Christ doth witness, without the grain of wheat
Die in the ground, encrease maist thou none get.

And in likewise without the matter putrifie,
It may in no wise truly alterate,
Neither thy elements may be divided kindly
Nor the conjunction of them perfectly celebrate;
That thy labour therefore be not frustrate,
The privitie of our putrifying well understand
Or ever thou take this work in hand.

And Putrifaction may thus defined be,
After philosophers sayings, to be of bodies the slaying;
And in our compound a division of things three
The killed bodies into corruption forth leading,
And after into regeneration them abling,
For things being in the earth, without doubt
Be engendered of rotation of the heavens about.

And therefore like as I have said before
Thine elements commixt and wisely cocquate
Thou keep in temperate heat eschewing evermore,
That they by violent heat be not incinerate,
To powder dry improfitably rubificate,
But into powder black as a crows bill,
With heat of balne or else of our dunghill.

Until the time that nights be passed ninety,
In moist heat keep them for anything,
Soon after by blackness thou shalt espie
That they draw fast to putrifying,
Which thou shalt after many colours bring
To perfect whiteness by patience easily,
And so thy seed in his nature shall multiplie.

Make each the other then to halse and kiss,
And like as children to play them up and down,
And when their shirts are filled with piss,
Then let the woman to wash be bowne,
Which oft for faintness will in a swoon,
And die at the last with her children all,
And go to purgatorie to purge their filth original.

When they be there, by little and little increase
Their pains, by heat, aye more and more,
The fire from them let never increase,
And so that thy furnace be surely apt therefore,

Which wise men call an Athenore,
Conserving heat required most temperatelie,
By which thy matter doth kindly putrifie.

Of this principle speaketh Sapient Guido,
And saith by rotting dyeth the compound corporal
And then after Morien and other more,
Upriseth again regenerate simple and spiritual,
And were not heat and moisture continual,
Sperm in the womb might have more abiding
And so there should no fruit thereof upspring.

Therefore at the beginning our stone thou take,
And bury each one in other within their grave.
Then equally betwixt them a marriage make
To ligge together six weeks let them have,
Their seed conceived, kindly to nourish and save,
From the ground of their grave not rising that while,
Which secret point doth many a one beguile.

This time of conception with easy heat abide,
The blackness shewing shall tell thee when they dye,
For they together like liquid pitch that tide,
Shall swell and burble, settle and putrifie,
Shining colours therein thou shalt espie,
Like to the rainbow marveilous to sight,
The water then beginneth to dry upright.

For moist bodies heat working temperate,

Ingendreth blackness, first of all which is
Of kindly conjunction the token assignate.
And of true putrifying; remember this
For then perfectly to alter thou canst not miss
For thus be the gate thou must come in
The light of Paradise in whiteness if thou wilt win.

For first the Sun in his uprising obscurate
Shall be, and pass the waters of Noah's flood
On earth which was an hundred days continuate
And fifty away ere all these waters yood;
Right so our waters (as wise men understood)
Shall pass, that thou with David may say
Albierunt in sicco flumina; bear this away.

Soon after that Noah planted his vineyard,
Which royally flourished, and brought forth grapes anon
After which space thou shalt not be afeared,
For in likewise shall follow the flourishing of our stone,
And soon after that days be gone
Thou shalt have grapes ripe as Rubie red
Which is our Adrop, our Vcifer[1], and our red lead.

For like as Souls after pains transitorie
Be brought to Paradise where ever is joyfull life,

---

[1] Lucifer?

So shall our stone (after his darkness in Purgatorie)
Be purged, and joined in Elements without strife,
Rejoice the whiteness and beautie of his wife,
And pass from darkness of Purgatorie to light
Of Paradise, in whiteness Elixer of great might.

And that thou maist the rather to Putrefaction win,
This example thou take to thee for a true conclusion,
For all the secret of Putrifaction resteth therein;
The heart of Oak that hath of water continual infusion
Will not soon putrifie, I tell thee without delusion;
For though it in water lay 100 years and more,
Yet shouldst thou find it sound as ere it was before.

But and thou keep it sometime wet and sometime dry
As thou maist see in timber by usual experiment
By process of time that oak shall putrifie;
And so even likewise according to our intent,
Sometimes our tree must with the sun be brent,
And then with water we must it keele,
That by this means of rotting we may bring it weele.

For now in wet, and now again in dry
And now in heat, and now again in cold
To be, shall cause it soon to putrifie,
And so shalt thou bring rotting thy gold;
Intreat thy bodies as I have thee told,
And in thy putrifying, with heat be not too swift,
Least in the Ashes thou seek after thy thrift.

Therefore thy water out of the earth thou draw,
And make the soul therewith for to ascend,
Then down again into the earth it throw,
That they oft times so ascend and descend,
From violent heat and sudden cold defend
Thy glass, and make thy fire so temperate
That by the sides the matter be not vitrificate.

And be thou wise in choosing of the matter,
Meddle with no salts, sulphurs, nor mean minerals;
For whatsoever any worker to thee doth clatter,
Our Sulphur and our Mercury been only in metals
Which oils and waters some men them calls,
Fowls and birds, with other names many one
Because that fools should never know our stone.

For of this world our stone is called the ferment
Which moved by craft as nature doth require
In his increase shall be full opulent,
And multiply his kind after thine own desire
Therefore if God vouchsafe thee to inspire,
To know the truth, and fancies to eschew
Like unto thee in riches shall be but few.

But many men be moved to work after their fantasie
In many subjects in which be tinctures gay;
Both white and red divided manually
To sight, but in the fire they fly away;

Such break pots and glasses day by day,
Enpoisoning themselves and loosing their sights
With odours, smokes, and watching up by nights.

Their clothes be baudy and worn thread bare,
Men may them smell for multipliers where they go
To fill their fingers with corrosives they do not spare
Their eyes be bleared, their cheeks lean and blow
And thus for had I wist they suffer loss and woe;
And such when they have lost that was in their purse,
Then do they chide, and Philosophers sore do curse.

To see their houses it is a noble sport,
What furnaces, what glasses there be of divers shapes
What salts, what powders, what oils, water sort.
How eloquently de materia prima their tongues do clap
And yet to find the truth they have no hap;
Of our mercury they meddle and of our sulphur vive
Where they dote, and more and more untrue.

For all the while they have Philosophers been,
Yet could they never know what was our stone.
Some sought it in dung, in urine, some in wine,
Some in star slime (for thing it is but one),
In blood, in eggs; some till their thrift was gone,
Dividing elements, and breaking many a pot,
Sheards multiplying, but yet they hit it not.

They talk of the Red man and his white wife,
That is a special thing and of the Elixirs two,
Of the Quintessence, and of the Elixir of life,
Of honey, celidonie, and of secondines also,
These they divide into elements, with others more,
No multipliers, but philosophers called will they be,
Which natural philosophers did never read nor see.

This fellowship knoweth our stone right well
They think them richer than is the King,
They will him help, he shall not fail
France for to win a wondrous thing,
Thy holy Cross home they will bring,
And if the King were prisoner I take,
Right soon his ransom would they make.

A mervaile it is that Westminster Kerke,
To the which these Philosophers do much haunt,
Since they can so much riches werke
As they make boast of an avaunt,
Drinking dayly the wine a due taunt,
Is not made up perfectly at once,
For truly it lacketh yet many stones.

Fools do follow them at the tail,
Promoted to riches weening to be;
But will you hear, what worship and avail
They win in London that noble city?
With silver maces (as you may see)

Sergeants awaiteth on them each hourly,
So been they men of great honour.

Sergeants seek them from street to street,
Merchants and goldsmiths lay after them watch,
That well is him that with them may meet,
For great advantage that they do catch,
They hunt about as doth a bratch,
Weening to win so great treasure
That ever in riches they shall endure.

Some would catch their goods again,
And some more good would adventure,
Some for to have would be full fain
Of ten pounds one, I you ensure.
Some which have lent without measure
Their goods, and be with povertie clad,
To catch a noble, would be full glad.

But when seargents do them arrest,
Their partners be stuffed with Paris balls,
Or with signets of Saint Martins at the least;
But as for money it is pist against the walls;
Then be they led (as well them befalls)
To Newgate or Ludgate as I you tell,
Because they shall in safeguard dwell.

Where is my money become, saith one?
And where is mine saith he and he.

But will you hear how subtile they be anon
In answering, that they be excused be?
Saying, of our Elixers robbed be we,
Else might we have paid you all your gold,
Though it had been more by ten fold.

And then their creditors they flatter so,
Promising to work for them again
In right short space the Elixirs two.
Doting the merchants that they be fain
To let them go, but ever in vain;
They work so long, till at the last
They be again in prison cast.

If any of them ask, why they be not rich?
They say they can make fine gold of tin
But he, (say they) may surely swim the ditch
Which is upholden by the chin;
We have no stock, therefore may we not win
Which if we had, we would soon work
Enough to finish up Westminster Kerk.

And some of them be so devout,
They will not swell out of that place;
For there they may withouten doubt
Do what them list to their solace,
The Archdeacon is so full of grace,
That if they bless him with their cross
He forceth little of other mens loss.

And when they then sit at the wine,
These monks they say have many a pound,
Would God (saith one) that some were mine,
Yet care away, let the cup go round;
Drink on saith another, the mean is found,
I ani a master of that Art
I warrant us we shall have part.

Such causeth Monks evil to doone,
To waste their wages through their dotage,
Some bringeth a mazer, some a spoon.
Their Philosophers giveth them such comage,
Behighting them winning with domage
A pound for a pennie at the least again;
And so fair promises make fools fame.

A royal medicine one upon twelve,
They promise them thereof to have
Which they could never for themselves
Yet bring about, so God me save;
Beware such philosophers no man deprave
Which help these monks to riches so
In thread bare coats that they must go.

The Abbot ought well to cherish this companie,
For they can teach his monks to live in povertie,
And to go cloathed in moneyed religioustie,
As did Saint Bennet, eschuing superfluities,

Easing them also of the ponderositie
Of their purses, with pounds so aggravate
Which by philosophie be now alleviate.

Lo who so medleth with this rich companie,
Great boast of their winning they may make;
For they shall reap as much by their philosophy
As they of the tail of an ape, can take;
Beware therefore for Jesus sake
A meddle with no thing of great cost
For if thou do, it is bust lost.

These Philosophers (of which I spake before)
Meddle and blunder with many a thing,
Running in errors ever more and more
For lack of true understanding;
But like must like always forth bring
So hath God ordained in every kind,
Would Jesus they would bear this in mind.

Weene they of a nettle to have a rose,
Or of an elder to have an apple sweet;
Alsa, that wise men their goods should loose,
Trusting such laurels when they them meet
Which say our stone is trodden under feet,
And maketh them vile things to distill,
Till all their gouses with stench they fill.

Some of them never learned a word at schools,

Should such by reason understand Philosophers?
Be they Philosophers? Nay, they be fools;
For their works prove them unwittie,
Meddle not with them, if thou be happie,
Least with their flatterie they do thee till
That thou agree unto their will.

Spend not thy money away in waste,
Give not to every spirit credence,
But first examine, grope and taste;
And as thou proovest, so put thy confidence,
But ever beware of great expence;
And if the philosopher do live vertuously,
The better thou maist trust his philosophy.

Prove him first, and him appose
Of all the secrets of our stone;
Which if he know not, thou need not to lose,
Meddle thou no further, but let him gone,
Make he never so perious a move;
For then the Fox can fag and fain,
When he would to his prey attain.

If he can answer as a clark,
How be it he hath nor proved indeed,
And thou then help him to his work;
If he be virtuous I hold it meed,
For he will thee quit if ever he speed
And thou shalt know by a little anon,

If he have knowledge of our stone.

One thing, one glass, one furnace, and no more.
Behold this principle if he do take,
And if he do not then let him go,
For he shall thee no rich man make
Timely it is better thou him forsake
Than after with loss and variance
And other manner of displeasance.

But if God fortune thee to have
This science by doctrine which I have told,
Discover it not whosoever it crave
For favour, fear, silver, or gold,
Be no oppressor, lecher nor boaster bold;
Serve thy God, and help the poor among,
If thou this life list to continue long.

Unto thyself thy secrets ever keep
From sinners, which have not God in dread
But will thee cast in prison deep,
Till thou them teach to do it indeed,
Then slander on thee shall spring and spread,
That thou doest coin then will they say,
And so undo thee for ever and aye.

And if thou teach them this cunning,
Then sinfull loving for to maintain,
In Hell therefore shall be thy wooning,

For God of thee and them will take disdaine,
As thou nought couldst therefore thee fame,
That bodie and soul thou maist both save,
And here in peace thy living to have.

Now in this chapter I have thee taught,
How thou thy bodies must putrifie,
And so to guide thee that thou be not caught
And put to durance loss or villaine,
My doctrine therefore remember wittely,
And pass forth towards the sixth Gate,
For this the fifth is triumphate.

The End of the Fifth Gate.

Of Congelation.

The Sixth Gate.

Of congelation I need not much to write,
But what it is, I will to thee declare;
It is of soft things induration of colour white,
And confixation of spirits which flying are;
How to congeal, he needth not much to care,
For Elements will knit together soon
So that Putrifaction be kindly done.

But Congelations be made in divers wise
Of spirits and bodies dissolved into water clear
Of salts also dissolved twice or thrice
And then congealed into a fluxible matter;
Of such congealing fools fast do clatter;
And some dissolveth dividing manually
Elements, them after congealing to powder drie.

But such congealing is not to our desire,
For unto ours it is contrarious.
Our congelation dreadeth not the fire;
For it must ever stand in it unctuous,
And it is also a tincture so bounteous,
Which in the air congealed will not relent
To water, for then our work were spent.

Moreover congeale not into so hard a stone
As glass or crystal, which melteth by fusion
But so that it like wax will melt alone
Withouten blast; and beware of delusion,
For such congealing accordeth not to our conclusion
As will not flow, but run to water again
Like salt congealed, then laboureth thou in vain.

Which congelation availeth us not a deale,
It longeth to multipliers; congealing vulgarly,
If thou therefore list to do well
(Sith the medicine shall never flow kindly,
Neither congeal, without thou first it putrifie)
First purge, and then fix the elements of our stone
Till they together congeale and flow anon.

For when thy matter is made perfectly white,
Then will the spirit with the bodie congealed be:
But of that time thou maist have long respite
Or it congeale like pearls in sight of thee,
Such congealation be thou glad to see
And after like grains red as blood,
Richer than any worldly good.

The earthly grossness therefore first mortified,
In moisture blackness ingendered is;
This principle may not be denied,
For natural Philosophers so sayne the wis,
Which had, of whiteness thou maist not miss

And into whiteness if thou congeale at once,
Then hast thou a stone most precious of all stones.

And by the dry like as the moist did putrifie,
Which caused in colour blackness to appear,
Right so the moist congealed by the dry,
Ingendreth whiteness shinning by night full clear,
And driness proceedeth as whiteth the matter,
Like as in blackness moisture doth him show
By colours variant always new and new.

The cause of all this is heat most temperate,
Working and moving the matter continually,
And thereby also the matter is alterate,
Both inward and outward substancially,
Not as do fools to fight sophistically;
But in every part all fire to indure
Fluxible, fixt, stable in tincture.

As physick determineth of each digestion,
First done in the stomach in which is driness,
Causing whiteness without question,
Like as the second digestion causeth redness
Complete in the liver by heat in temperateness,
Right so our stone by driness and by heat
Digested is to white and red compleate.

But here thou must another secret know,
How the Philosophers child in the air is borne,

Busie thee not too fast at the coal to blow,
And take this neither for mock nor scorn,
But trust me truly, else is all thy work forlorn,
Without thine earth with water reunited be
Our true congealing shalt thou never see.

A soul it is betwixt heaven and earth being,
Arising from earth as air with water pure,
And causing life in every lively thing
Incessable running upon our fourfold nature
Enforcing to better him with all his cure,
Which air is the fire of our Philosophie;
Named now oil, now water mysticallie.

And this means air which oil or water we call
Our fire, our ointment, our spirit, and our stone,
In which one thing we ground our wisedomes all,
Goeth neither in nor out alone,
Nor the fire but the water alone;
First it out leadeth, and after it bringeth it in,
As water with water which will not lightly twin.

And so may water only our water meene,
Which moving causeth both death and life
And water to water doth kindly cleeve
Without repugnance or any strife,
Which water to fools is nothing rife,
Being of the kind withouten doubt
Of the spirit, called water and leader out.

And water is the secret and life of every thing,
That is of substance in this world I found,
For of water each thing hath his beginning,
As showeth in women when they shall be unbound
By water, which passeth before it all be found
Called Alvien, first from them running,
With greivous throwes before their childing.

And truly that is the cause most principal
Why Philosophers charge us to be patient,
Till time the water be dried to powder all
With nourishing heat, continual, not violent;
For qualities be contratious of every element,
Till after black in white be made an union
Of them for ever, congealed without division.

And furthermore, the preparation of this conversion;
From thing to thing, from one state to another,
Is done only by kindly and discreet operation
Of nature, as is of sperm within the mother
For sperm and heat, are as sister and brother,
Which be converted in themselves as nature can,
By action and passion at last to perfect man.

For as the bodily part by nature was combined
Into man, is such as the beginner was,
Which though it thus fro thing to thing was alterate
Not out of kind to mix with other kind did pass

And so our matter spermatical within our glass,
Within it self must turn from thing to thing,
By heat most temperate only it nourishing.

Another example natural I may tell thee,
How the substance of an egg by nature is wrought
Into a chicken not passing out of the shell,
A plainer example could I not have thought,
And their conversions be made till forth be brought
From state to state, the like by like in kind,
With nourishing heat only bear this in mind.

Another example here also thou maist read
Of vegetable things, taking consideration,
How everything groweth of his own seed
Through heat and moisture, by natural operation,
And therefore minerals be nourished by ministration
Of moisture radical, which there beginning was,
Not passing their kind within one glass.

There we them turn from thing to thing again,
Into their mother the water when they go;
Which principle unknown, thou labourest in vain,
Then all is sperm; and things there be no more
But kind with kind in number two,
Male and female, agent and patient,
Within the matrix of the Earth most orient.

And these be turned by heat from thing to thing

Within one glass, and so from state to state,
Until the time that nature doth them bring
Into one substance of the water regenerate;
And so the sperm with his kind is alterate,
Able in likeness his kind to multiply,
As doth in kind all other things naturally.

In the time of this said process natural,
While that the sperm conceived is growing,
The substance is nourished with his own menstruale,
Which water only but out of the earth did spring,
Whose colour is green in the first showing;
And from that time the sun hideth his light,
Taking his course throughout the North by night.

The said menstruall is (I say to thee in councel)
The blood of our Green Lyon and not of vitriol,
Dame Venus can the truth of this thee tell
At the beginning, to councel if thou her call,
This secret is hid by Philosophers great and small,
Which blood drawn out of the Green Lyon,
For lack of heat had not perfect digestion.

But this blood called our secret menstraull,
Wherewith our sperm is nourished temperately
When it is turned into the feces corporal,
And so become white perfectly and very dry,
Congealed and fixed into his own bodie,
Then decoct blood to sight it may well seem,

Of this work named the milk white Dyademe.

Understand now that our fine water thus acuate,
Is called our menstruall water, wherein,
Our earth is loosed and naturally calcinate,
By Congelation that they may never twinne,
But yet to congeal more water thou may not linne;
Into three parts of the acuate water sayed afore
With the fourth part of the earth congealed no more.

Unto that substance therefore congelate,
The fourth part put of water christalline,
And make them then together to be dispousate,
By Congelation into a miner metalline,
Which like a sword new slipped will shine,
After the blackness which first will shew,
The fourth part then give it of water new.

Imbibitions many it must yet have,
Give it the second, and after the third also,
The said proportion keeping in thy witt,
Then to another the fourth time look thou go,
The fifth time and the sixth pass not therefore,
But put two parts at each time of them three,
And at the seventh time five parts must there be.

When thou hast made seven times Imbibition,
Again then must thou turn about thy wheel,
And putrifie all that matter without addition.

First blackness abiding if thou wilt do well,
Then into whiteness congeal it up each deele,
And after by redness into the South ascend,
Then hast thou brought thy base unto an end.

Thus is thy water then divided into parts two,
With the first part the bodies be putrificate,
And to thine Imbibitions the second part must go,
With which thy matter is afterward demigrate
And soon upon easie decoction albificate,
Then is it named by Philosophers our starry stone,
Bring that to redness then is the sixth gate won.

End of Sixth Gate.

Of Cibation.

The Seventh Gate.

Now of Cibation I turn my pen to write,
Sith it must here the Seventh place occupie,
But in few words it will be expedite,
Take heed therefore, and understand me wittelie,
Cibation is called a feeding of our matter drie,
With milk and meat, which moderately thou do,
Until it be brought the third order unto.

But give it never so much, that thou it glut,
Beware of dripsie, and also of Noahs flood:
By little and little therefore thou to it put
Of meat and drink, as seems to do it good,
That watery humours not overgrow the blood,
To drink therefore let it be measured so,
That kindly appetite thou never quench it fro.

For if it drink too much, then must it have
A vomit or else it will be sick too long,
From the dropsie therefore thy womb thou save,
And from the flux, or else it will be wrong,
But rather let it thirst for drink among,
Than thou shouldst give it over much at once,
Which must in youth be dieted for the nonce.

And if thou diet it (as nature doth require)

Moderately, till time that it be grown to age,
From cold it keeping, and nourishing with moist fire,
Then shall it grow, and wax full of courage,
And do to thee both pleasure and advantage;
For he shall make dark bodies whole and bright,
Cleansing their leprosy through his might.

Three times must thou turn about thy wheel,
Still keeping the rule of the said Cibation,
And then as soon as it the fire doth feel,
Like wax it will be readie unto liquation;
For I have told thee the dietorie most convenient,
After thine Elements be made equipolent.

And also how to whiteness thou shalt bring thy gold,
Most like in figure to leaves of hawthorne tree
Called Magnesia, afore as I have told,
And our white sulphur without combustibilitie,
Which from the fire away will never flie,
And thus the seventh Gate (as you desired)
In the uprising of the Sun is conquered.

End of Seventh Gate

Of Sublimation.

The Eighth Gate.

Here of our Sublimation a word of two
I have to speak, which the eighth gate is,
Fools do sublime, but sublime thou not so,
For we sublime not as they do y wis.
To sublime truly therefore thou shalt not miss,
If thou canst make thy bodies first spiritual,
And then thy spirits (as I have taught thee) corporall.

Some do mercurie from vitriol and salt sublime,
And other spirits from scales of iron and steel,
From egg-shells calcined, and from quick lime.
And in their manner yet sublime they right well;
But such subliming accordeth never a deele
To our intents, for we sublime not so,
To true subliming therefore, now will I go.

In Sublimation first beware of one thing,
That thou sublime to the top of the vessel not;
For without violence thou shalt it not down bring
Again, but there it will abide and dwell.
So it rejoiceth with refrigeration I thee tell,
Keep it therefore with temperature heat down
Full forty days, till it wax black and brown.

For then the soul beginneth to come out
From his own veyness, for all that subtil is
Will with the spirit ascend withouten doubt,
Bear in thy mind therefore, and think on this,
How here eclipsed been thy bodies,
As thou do putrifie subliming more and more
Into water, until they be all up above.

And thus their venome when they have spued out
Into the water then black it doeth appear,
Becoming spiritual each deale without doubt,
Subliming easily in our manner,
Into the water, which doth him bear;
For in the air our child must thus be bore
Of the water again, as I have said before.

But when these two by sublimation continual
Be laboured so with heat both moist and temperate,
That all is white and purely made spiritual,
Then heaven upon earth must be reiterate,
Until the soul with the bodie be incorporate
That earth become all that before was heaven
Which will be done in sublimations seven.

And Sublimations we make for causes three,
The first cause is, to make the bodie spiritual;
The second is, that the spirit may corporal be,
And become fixt with it, and consubstantial;

The third cause is, that from his filthy original
He may be cleansed, and his saltness sulphurious
May be mingled in him, which is infectious.

Then when they thus together depured be,
They will sublime up whiter than snow.
That sight will greatly comfort thee;
For then anon perfectly thou shalt know
The spirits shall so adowne y throwe,
That this eighth gate shall be to thee unlocked
Out of the which many be shut and mocked.

End of Eighth Gate.

Of Fermentation.

The Ninth Gate.

True Fermentation few workers understand,
That secret therefore I will expound to thee,
I travelled through truly many a land,
Ere ever I might find any that would tell it me,
Yet as God would, evermore blessed be he,
At the last I came to the knowledge thereof,
Take heed therefore what I thereof do write.

Fermentation in divers manners be done
By which our medicine must be perpetuate
Into clear water; Some looseth Sun and Moon,
And with their medicines make them to be congelate,
Which in the fire when they be examinate
May not abide, nor alter with complement;
For such Fermenting is not to our intent.

But yet more kindly so me other men doone,
Fermenting their medicines in this wise,
In mercurie dissolving both Sun and Moon,
Till time with spirit they will arise,
Subliming them together twice or thrice,
Then Fermentation therewith they make,
That is away, but yet we it forsake.

Some other there be which have more hap,

To touch the truth in part of fermenting,
They amalgame their bodies with mercurie like pap,
Then thereupon their medicines relenting
These of our secrets have some hinting.
But not the truth with perfect compliment,
Because they neither putrifie, nor alter their ferment.

That point therefore I will disclose unto thee,
Look how thou didst with thine imperfect bodie,
Do so with thy perfect bodies in each degree,
That is to say, first thou them putrifie,
Their former qualities destroying utterly,
For this is wholly to our intent,
That first thou alter before ferment.

To thy compound make ferment the fourth part
Which ferments being only of Sun and Moon;
If thou therefore be master of this Art,
Thy Fermentation let thus be done,
Fix water and earth together soon,
And when thy medicine as wax doth flow,
Then upon malgames look thou it throwe.

And when all that together is mixed,
Above the glass well closed make thy fire
And so continue it till all be fixed
And well fermented to thy desire,
Then make projection after thy pleasure,

For that is medicine each deale perfite,
Thus must thou ferment both red and white.

For like as flour of wheat made into paste
Requireth ferment, which leaven we call
Of bread, that it may have the kindly taste
And become food to man and woman cordial
Right so thy medicine ferment thou shall
That it may taste of the ferment pure
At all assays for ever to endure.

And understand that there be ferments three,
Two be of bodies of nature clean,
Which must be altered as I have told thee;
The third most secret of which I mean,
Is the first earth of his water Green;
And therefore when the Lion doth thirst,
Make him drink till his belly burst.

Of this a question if I should mouve,
And ask of workers, what is this thing?
Anon thereby I should them prove,
If they had knowledge of our fermenting;
For many a man speaketh with wondering,
Of Robinhood and of his bow,
Which never shot therein I trowe.

For fermentation true as I thee tell,
Is of the soul with bodies incorporation,

Restoring to it the kindly smell,
With taste and colour by natural conspissation,
bf things dissevered , a due reintegration,
Whereby the body of the spirit taketh impression,
That either the other may help to have ingression.

For like as bodies in their compaction corporall,
May not show out their qualities effectually,
Until the time that they become spiritual,
No more may spirits abide with bodies stedfastly,
Till they with them be confixate proportionally
For then the body teacheth the spirit to suffer fire,
And the spirit the body to enter to thy desire.

Therefore thy gold with gold thou must ferment,
With his own water thy earth cleansed I mean,
Nought else to say but element with wlement,
The spirit of life only going between,
For like as an adamant as thou hast seen
Draweth iron to him, so doth our earth by kind
Draw down to him his soul born up with wind.

With wind therefore the soul lead out and in,
Mingle gold with gold, that is for to say,
Make element with element together rin
Till time all fire they suffer may,
For earth is ferment, withouten nay,
To water, and water the earth unto,
Our Fermentation in this wise must be do.

Earth is gold, and so is the soul also
Not common, but ours thus Elementate,
And yet thereto the sun must go,
That by our wheel it may be alterate;
For so to ferment it must be preparate,
That it profoundly may joined be,
With other natures as I said to thee.

And whatsoever I have here said of gold,
The same of silver I will thou understand,
That thou them putrifie and alter (as I have told)
Ere thou thy medicine to ferment take in hand;
Forsooth I could never find him in England
Which in this wise to ferment could we teach
Without error, by practice only speech.

Now of this chapter needeth to treat no more,
Sith I intend prolixitie to eschew;
Remember well my words therefore,
Which thou shalt prove by practice true,
And sun and moon look thou renew,
That they may hold of the first nature,
Then shall their tincture evermore endure.

And yet a way there is most excellent,
Belonging unto another working,
A water we make most redolent,
All bodies to oil wherewith we bring,

With which our medicine we make flowing,
A quintessence this water we call,
In man which healeth diseases all.

But with thy base, after my doctrine prepare,
Which is our calx this must be done,
For when our bodies de so calcinate,
That water will to oil dissolve them soon,
Make thou therefore oil both of sun and moon,
Which is ferment most fragrant for to smell
And so the ninth Gate is conquered of this castle.

End of Ninth Gate.

Of Exaltation.

The Tenth Gate.

Proceed we now to the chapter of Exaltation,
Of which truly thou must have knowledge pure,
But little it is different from Sublimation,
If thou conceive it right I you ensure,
Hereto accordeth the holy Scripture,
Christ saying thus, it I exalted be,
Then shall I draw all things unto me.

Our medicine if we exalt right so,
It shall thereby nobilitate,
That must be done in manners two,
From time the parties be dispousate,
Which must be crucified and examinate,
And then contumulate both man and wife,
And after reunited by the spirit of life.

Then up to heaven they must exalted be,
There to be in bodie and soul glorified (glorificate)
For thou must bring them to such subtiltie,
That the ascend together to be intronizate,
In clouds of clearness to Angels consociate,
Then shall they draw as thou shalt see,
All other bodies to their own dignitee.

If thou therefore the bodies wilt exalt,
First with the spirit of life thou them augment,
Till time the earth be well subtilizate,
By natural rectifying of every element,
Them up exalting into the firmament,
Then much more precious shall they be than gold,
Because of the quintessence which they do hold.

For when the cold hath overcome the heat,
Then into water the air shall turned be,
And so two contraries together shall meet
Till either with other right well agree,
So into air the water as I tell thee,
When heat, of cold hath got domination,
Shall be converted cast of our circulation.

And of the air then fire have thou shall
By loosing, putrefying and subliming,
And fire thou hast of the earth material,
Thine elements thus by craft dissevering,
Most especially thine earth well calcining.
And when they be each one made pure
Then do they hold all of the first nature.

On this wise therefore make them be circulate,
Each into other exalting by and by,
And all in this one glass surely sigillate,
Not with thine hands, but as I teach thee naturally,
Fire into water then turn first hardly,

For fire is in air, which is in water existent
And this conversion accordeth to our intent.

Then further more turn on thy wheel,
That into earth the air converted be,
Which will be done also right well
For air is in water being in earth trust me,
The water into fire contrarious in her qualitie,
Soon turn thou mayst for water in earth is,
Which is in fire, conversion true is this.

The wheel is now near turned about
Into air turn earth which is the proper nest,
Of other elements there is no doubt,
For earth is fire is, which in air taketh rest,
This circulation begin thou in the West,
Then into the fourth, till they exalted be,
Proceed duely, as in thy figure I have taught thee.

In which process clearly thou mayst see
From one extreme how to another thou mayst not go
But by a mean, since they in qualities contrarious be,
And reason will forsooth, that it be so,
As heat into cold, with other contraries mo,
Without their means as moist to heat and cold
Examples sufficient before this I have told.

Thus have I taught thee how to make
Of all thine elements a perfect circulation,

And at thy figure example to take,
How thou shalt make this foresaid Exaltation,
And of thy medicine in the Elements true graduation
Till it be brought to a gueneritie temperate,
And then thou hast conquered the tenth Gate.

                The End of the Tenth Gate.

The Eleventh Gate.

Of Multiplication.

Multiplication now to declare I proceed
Which is by Philosophers in this wise defined
Augmentation it is of the Elixer indeed,
In goodness and quantitie both for white and red,
Multiplication is therefore as they do write,
That thing that doth augment medicines in each degree,
In colour, in odour, in virtue and also in quantitie.

And why thou mayst this medicine multiplie
Infinetely forsooth the cause is this,
For it is fire, which kindled will never die,
Dwelling with thee, as fire doth in houses,
Of which one spark may make more fire I wis,
As musk in pigments and other spices mo
In virtue multiplied, and our medicine right so.

So he is rich which fire hath less or more,
Because he may so hugely it multiply,
And right so rich is he which any part hath in store,
Of our Elixers which be augmentable infinitely,
One way if thou dissolve our powders dry,
And make often time of them congelation,
There of in goodness then makest thou augmentation.

The second way both in goodness and quantitie
It multiplyeth by iterate Fermentation,
As in that chapter I showed plainly to thee,
By divers manners of natural operation,
And also in the chapter of our Cibation,
Where thou mayst know how thou shalt multiplie,
Thy medicine with mercurie infinetely.

But thou wilt both loose and eke ferment,
Both more in quantitie and better will it be,
And in such wise thou mayst it soon augment,
That in thy glass it will grow like a tree,
The tree of Hermes named seemly to see
Of which one pippin a thousand will multiplie,
If thou canst make thy projection wittely,

And like as saffron when it is pulverized
By little and little if it with liquor be
Tempered, and then with much more liquor dilate,
Teyneth much more of liquor in quantitie,
This being whole in his grosse nature; so shalt thou see
That our Elixir, the more it is made thinne
The further in tincture it fastly will rinne.

Keep in thy fire therefore both even and morrow
From house to house that thou had not to rinne
Among thy neighbours thy fire to seek or borrow
The more thou keepest, the more shalt thou win

Multiplying it always more and more thy glass within,
By feeding with mercurie unto thy lives end,
So shalt thou have more than thou needest to spend.

This matter is plain I will no more
Write thereof, let reason thee guide,
Be never the bolder to sin therefore,
But serve thy good the better in each tide;
And while that thou shalt in this life abide,
Bear this in mind, forget not I thee pray
As thou shalt appear before God at dooms day,

His own great gifts therefore and his treasure,
Dispose thou virtuously, helping the poor at need,
That in this world thou mayst to thee procure,
Mercy and Grace with heavenly bliss to to meede,
And pray to God devoutly that he thee lead,
In at the twelfth Gate, as he can best,
Soon after then thou shalt end thy conquest.

End of the Eleventh Gate.

Of Projection.

The Twelfth Gate.

In projection it shall be proved if our practice be profitable,
Of which it behoveth me the secrets here to move,
Therefore if thy tincture be sure and not variable,
By a little of thy medicine thus mayst thou prove,
With mettle, or with Mercury as pitch it will cleave,
And teyne in projection all fires to abide,
As soon it will enter and spread him full wide.

But many by ignorance do marr that they make,
When on metals uncleansed Projection they make,
For because of corruption their tinctures must fade,
Which they would not away first from the body take,
Which after projection be brittle blue and black,
That thy tincture may ever more last,
First upon ferment thy medicine see thou cast.

Then brittle as glass will thy ferment be,
Upon bodies cleansed and made pure,
Cast that brittle substance and soon shalt thou see
That they shall be curiously coloured with tincture,
With all assayes for ever shall endure
But profitable projection perfectly to make,
At the Psalms of the Psalter example thou take.

On Fundamenta cast first this psalm-Nunc dimitis,
Upon Verba Mea, then cast Fundamenta beline,
Then Verba upon diligam, conceive me with thy wits,
And diligam upon attendite, if thou list to thrive,
Thus make thou projections, three, four and five,
Till the tincture of thy medicine begin to decrease,
And then it is time of Projection to cease.

By this mistie talking I mean nothing else
But that thou must cast first the less on the more,
Encreasing aye the number as wise men thee tells
And keep thou this secret unto thy self in store,
Be covetous of cunning it is no burden sore,
For he that joyneth not the Elixer with bodies made clean
He wotteth not surely what projection doth mean.

Ten if thou multiply first into ten,
One hundred that number make sickerly,
If one hundred into an hundred be multiplied then,
Ten thousand is that number if thou count it wittely,
Then into as much more ten thousand to multiplie,
It is a thousand thousand, which multiplied Y wis,
Into as much more a hundreth millions is.

That hundreth millions being multiplied likewise,
Into ten thousand millions, as I to thee do say,
Maketh so great a number I wot not what it is,
Thy number in projection thus multiply alway.

Now Child of thy courtesie for me that thou pray,
Sith I have told thee our secrets all and some,
To the which I beseech God by grace thou mayst me.

Now hast thou conquered these Gates twelve,
And all the castle thou holdest at thy will;
Keep thy secrets in store to thyself,
And the commandments of God Look thou fulfill,
In fire see thou continue thy glass still
And multiply thy medicines aye more and more,
For wise men do say that store is no sore.

End of the Twelve Gates.

Ricapitulatlo Totius Opens Proedicti.

For to bring this treatise to a final end,
And briefly here to conclude these secrets all,
Diligently look thou, and to thy figure attend,
Which doth in it contain these secrets great and small,
And if thou it conceive, both theoritical and practical,
By figures and colours, by scripture plain,
It wittily conceived, thou may'st not work in vain.

Consider first the latitude of this precious stone,
Beginning in the first side noted in the West,
Where the red man and the white woman be made one,
Spoused with the spirit of life to live in rest,
Earth and water equally proportionate, that is best,
And one of the earth is good, and of the spirit three,
Which twelve to four also of the earth may be.

Three of the wife, and one of the man thou take,
And the less of the spirit in this dispousation,
The rather thy Calcination for certain shalt thou make,
Then forth into the North proceed by obscuration
Of the red man and his white wife, called Eclipsation,
Loosing them and altering them betwixt winter and vere,
Into water turning earth, dark and nothing clear.

From thence by colours many one into the East ascend,
Then shall the Moon be full appearing by day-light,
Then is she passed purgatorie, and her course at an end,
There is the uprising of the Sun appearing bright,
There is summer after vere, and day after night;
Then earth and water which were black, be turned to air,
And clouds of darkness overblown, and all appeareth fair.

And as in the West was the beginning of thy practice,
And the North the perfect mean of profound alteration;
So in the East after them the beginning of speculation is;
But of this course up in the South the Sun maketh consumation,
Their bin the elements turned into fire by circulation;
Then to win to thy desire thou needst not be in doubt,
For the wheel of our philosophie thou hast turned about.

But yet about again two times turn thy wheel,
In which bin comprehended all the secrets of our philosophy
In Chapters 12, made plain to thee, if thou conceive them well,

And all the secrets by and by of our lower Astronomy
How thou shalt calcine bodies, dissolve divide and putnify,
With perfect knowledge of all the poles which in our Heaven been,
Shining with colours inexplicable, never were gayer seen.

And thus our secret conclusion know withouten fails,
Our red man teineth not, nor his wife, till they teined be,
Therefore if thou list thy self by this craft to avail,
The altitude of the bodies hide, and show out their profunditie
In every of thy materials destroying the first qualitie,
And secondary qualities more glorious in them repair anon,
And in one glass, and without rule, four natures turn to one.

Pale and black with false citrine, imperfect white and red,
The peacocks feathers in colours gay, the Rainbow which shall overgo,
The spotted panther, the Lion Green, the Crows bill blue as lead,
These shall appear before thee perfect white and many

other more,
And after the perfect white, gray, false citrine also,
And after these, then shall appear the body red invariable,
Then hast thou a medicine of the third order of his own kind multiplicable.

Thou must divide thy white Elixir into parts two,
Before thou rubifie, and into glasses two let them be done,
If thou have for Sun and Moon thy elixir both do SO;
And into Mercury then multiply to great quantity soon,
And if thou had not at the beginning to fill a spoon,
Yet maist thou them so multiply both white and red
That if thou live a 1000 years, they shall stand thee in stead.

Have thou recourse to thy wheel therefore I counsel thee,
And studie him well to know in each chapter truly,
Meddle with no phantastical multipliers, but let them be,
Which will thee flatter feinning cunning in philosophy,
Do as I bid thee, then dissolve these foresaid bases wittilie,
And turn them into perfect oils with our true water ardent,
By circulation that must be done according to our

intent.

These oils will fix crude Mercury and convert bodies all
Into perfect Sun and Moon, when thou shalt make projection;
That oilie substance pure and fixt Raymond Lully did call
His Basiliske, of which he never made so plain detection;
Pray for me to God, that I may be one of his election,
And that he will for one of his, at doomsday me ken,
And grant me his bliss to reign with him for ever.

Amen.

Finis Recapitulationis.

The Epistle to King Edward the Fourth.

By the same Author.

1.   O Honourable Lord, and most victorious Knight,
With grace and virtue abundantly endued,
The safeguard of England, and maintainer of right;
That God you loveth, indeed he hath well shewed;
Wherefore I trust this land shall be renewed
With joy and riches, with charitie and peace,
So that old ranckors new understrewed,
Tempestuous troubles, and wretchedness shall cease.

2.   And therefore sith I see by tokens right evident,
That God you guideth, and how that you be vertuous,
Hating sin, and all such as be insolent,
How that also manslaughter to you is odious,
Upon the indgement also that you be piteous;
Me seenieth ruth it were but that you should live long;
For of your great fortune you are not presumptuous,
Nor vengeable of spirit to revenge you of each wrong.

3.   These considered with others in your most noble state,
Like as God knoweth, and people do witness bear,
So entirely me mooveth, that I must algate
Record the same, and therein be no flatterer;
And not that only, but also to write here

Unto your Highness, humbly to present
Great Secrets, which in far countries I did learn,
And which by grace to me most unworthy are lent.

4.   Once to your Lordship such things I did promise,
What time you did command to send unto me,
And sith that I wrote it in secret wise,
Unto your Grace from the Universitie,
Of Louaine, when God fortuned me by grace to see
Greater secrets and much more perfect,
Which only to you I will disclosed to be,
That is the Great Elixir both red and white.

5.   For like it you to trust that truly I have found,
The most perfect way and most secret Alchymie,
Which I will never truly for marke nor for pound
Make common but to you, and that conditionally.
That to yourself you shall keep it full secretly,
And use it as may be to Gods pleasure,
Else in time coming to God I should obey
For my discoursing of his secret treasure.

6.   Therefore be you well advised and with good deliberation,
For of this secret shall know no other creature,
But only you as I make faithfull protestation,
For all the time that here in life I shall endure,
Whereto I will your Lordship me ensure,
To my desire in this my oath for to agree,

Least I to me the wrath of God procure,
For such revealing of his great gift and privitie.

7.   If God fortune you by me to win this treasure,
Serve him devoutly with more lawde and thanking,
Praying his Godhead, in life that you may so endure,
His Gifts of grace, and fortune to use to his pleasing,
Most especially intending over all thing,
To your power and cunning his precepts ten
So to observe, that into no danger yourself you bring,
But that you in glory may see him hereafter, Amen.

8.   And yet moreover I will your Lordship to pardon ire,
For openly with pen I will it never it write,
But whensoever you list by practice you shall see,
By mouth also this precious secret, most of delight,
How may be made perfect Elixirs both red and white,
Plain unto your Lordship it shall declared be,
And if it please you, with easie expences and respite,
I will them work by grace of the Trinitie.

9.   But notwithstanding for peril that may befall,
If I dare not here plainly the knot unbind,
Yet in my writing I will not be so mistical,
But that to studie the true knowledge you may find,
How each thing is multiplied in its kind,
And how the likeness of bodies metalline be

transmutable
I will declare, that if you feel me in your mind
My writing you shall find true and no famed fable.

10. As philosophers in metheors do write,
The likeness of bodies metalline be not transmutable,
But after be added these words of more delight,
Without they be reduced to their beginning materiable,
Wherefore such bodies within nature be liquiable,
Mineral and metalline may be mercurizate,
Conceive you may this science is not opinionable,
But very true, by Raymond and others determinate.

11. In the said book the Philosophers speak also,
Therein if it please you Highness for to read,
Of divers sulphurs, and especially of two,
And of two mercuries joined to them indeed,
Whereby he doth true understanders lead
To knowledge of the principle which is only true,
Both red, most pure, and white, as I have espied,
Which be nevertheless found but of very few.

12. And these two things be best, be added anon,
For him that work the Alchymie to take
Our Gold and our Silver therewith to make all one,
Wherefore I say who will our pearl and Ruby make,
The said principles look he not forsake;
For at the beginning, if his principles be true,
And if so be by craft he can them also bake,

In the end truly his work he shall not rue.

13. But one great secret right needfull to be known,
That though the Philosophers speak plurally,
All is but one thing you may me well trowe,
In kind which is our base principally,
Whereof doth spring both white and red naturallly,
And yet the white must come first out of the red
Which thing is not wrought manually,
But naturally, craft helping out of our lead.

14. For all the parts of our most precious stone,
As I can prove, be co-essential and concrete,
Moreover there is no true principle but one,
Full long it was ere I therewith could meet,
Who can reduce him and knoweth his heat,
And only kind with kind can well redress
Till filthie original he cleansed from his seat,
He likely is to find our secrets more and less.

15. Therefore work kind only with his own kind,
And so your elements join that they not strive.
This point also for any bear in mind,
That passive natures you turn into active
Of water, fire, and wind of earth make bline[2]
And of the quadrangle make a figure round,
Then have thou the honey of thy bee-hive,

---

[2] Quickly or gladly

One ounce well worth one thousand pound.

16. The principal secret of secrets all,
Is true proportion which may not be behind,
Wherein I counsel thee be not superficial,
The true conclusion if thou thick to find,
Turn earth into water and water into wind,
Thereof make fire and beware of the flood
Of Noah, wherein many men are so blind
That by this science they get little good.

17. I counsel you eat and drink temperately,
And beware well that Iposarcha come not in place,
Neshe not your womb by drinking immoderately,
Lest you quench natural heat in little space,
The colour will tell appearing in your face,
Drink no more therefore then you may eat,
Walk up and down after an easie pace,
Cause not your bodie too sore to sweat.

18. With easie fire after moving when you sweat,
Warm your bodie and make it dry again,
By rivers and fountains walk after meat,
At morning time visit the high mountain,
That Phisick so bodden I reach certain,
So high the mountains yet do you not ascend
But that you may downwards your way have plain
And with your mantle from gold ye you defend.

19. Such labour is wholsome yor sweat for dry
With napkin, and after it see you take no cold,
For gross humours be purged by sweat kindly.
Use Diacameron then confect with perfect gold
Hermidocles for watery humours good I hold,
Use Ipericon perforat with milk of tinctural,
And Spermacete with red wine, when you wax old
And goats milk sod with wine nourisheth moisture radical.

20. But a good Physican who so intendeth to be,
Our lower Astronomie needth well to know
And after it need to learn well urine in a glass to see,
And if it need to be hansed the fire for to blow,
Then wittely it by divers ways for to throw
After the cause to make a medicine blive (quickly)
Truly telling the infirmities all on a row,
Who this can do by his Phisick is like to thrive.

21. We have our heaven incorruptible of the quintessence,
Ornate with signs, elements, and stars bright,
Which moistenth our earth by subtile influence;
And of it a secret sulphur hid from sight.
It fetcheth by virtue of his active might,
Like a bee fetcheth honey out of the flowers,
Which thing could do no other worldly wight.
Therefore to God be all glory and honour.

22. And like as ice to water doth relent,
When it was congealed by violence of cold,
When Phoebus yet shineth with his heat influent,
Even so to water mineral reduced is our gold,
As witnesseth plainly, Albert, Raymond, and Arnold,
By heat and moisture and by craft occasionate,
Which congelation of the spirits, be now I have told,
How our materials together must be proportionate.

23. At the dyers craft you may learn this science,
Beholding with with water how decoction they make
Upon a wode or madder easily and with patience,
Till tinctures do appear which when the cloth doth take,
Therein so fixed that they will never forsake
The cloth, for washing after they joined be,
Even so our tinctures with water of our lake,
We draw by boiling with the Ashes of Hermes tree.

24. Which tinctures when they by craft are made perfect,
So dyeth metals with colours aye permanent,
After the qualities of medicine, red or white,
That never away by any fire will be brent:
To this example if you take good tent,
Unto your purpose the rather you shall win,
And let your fire be easie, and not too fervent,
Where nature did leave off what time look you begin,

25. First calcine, and after that putrifie,
Dissolve, distill, sublime, discend, and fix,
With Aqua vitae oftentimes both wash and dry,
And make a marriage the body and spirit betwixt,
Which thus together naturellie if you can mix,
In loosing of the bodie the water congealed shall be,
Then shall the bodie die utterly of the flixe,
Bleeding and changing his colours, as you shall see.

26. The third day again to life he shall arise,
And devour birds, and beasts of the wilderness,
Crows, popinjaies, pies, peacocks, and mavies
The phoenix, with the eagle, and the griffon of fearfulness
The green lion with the red dragon, he shall distress,
With the white dragon, and the antelope, unicorn and panther,
With other beasts and birds both more and less
The Basiliske also, which almost each one doth fear.

27. In bus and nubibus he shall arise and descend,
Up to the moon, and sith up to the sun
Through the Ocean Sea, which round is withouten end,
Only shippen within a little glass tunne,
When he is there come, then is the masterie won,
About which journey, great goods you shall not spend,
And yet you shall be glad that ever it was begun,
Patiently if you list to your work attend.

28. For then both bodie and spirit, also both oil and water
Soul, and tincture, one thing both white and red,
After colours variable it containeth, whatsoever men clatter,
Which also is called after he hath once been dead,
And is revived, our markasite, our magnet and our lead,
Our Sulphur, our arsenick, and our true calx vive,
Our sun our moon, our ferment and our bread,
Our toad, our Basiliske our unknown bodie our man our wife,

29. Our bodie thus naturally by craft when he is renovate
Of the first order, is medicine called in our philosophy
Which oftentimes again must be Spiritualizate;
The round wheel turning of our Astronomie;
And so to the Elixir of spirits you must come; for why
Till the son of the fixed by the son of the fixer be overgone,
Elixir of bodies, nemed it is only,
And this sound secret point, deceaveth many an one.

30. This natural process by help of craft thus consummate,
Dissolveth Elixir spiritual in our unctious humiditie,

Then in Balneo Mariae together let them be circulate
Like new honey or oil, till perfectly they be thickened,
Then will that medicine heal all infirmitie,
And turn all metals to sun and moon perfectly
Thus you shall make the great Elixir and Aurum potabile,
By the Grace and Will of God, to whom be all honour and Glory.

        Amen Quod George Ripley

             Finis

GEORGE RIPLEY

MEDULLA ALCHYMIÆ

THE MARROW OF ALCHEMY

Written in Latin by George Ripley, Canon of Bridlington, which he sent out of Italy anno 1476. To the Arch-Bishop of York.

Translated into English and now revised and claused by:

WILLIAM SALMON

The Marrow of Alchemy

(Medulla Alchymie)

By George Ripley
1476

William Salmon, Translator

CHAPTER LXI

THE PREFACE TO THE ARCH-BISHOP OF YORK,

I. I Shall endeavour Sir, to explicate, open, and make plain to you, the Secrets of *Alchymie,* which I have attained to, by my Travels through *Italy;* and other Countries and Kingdoms for the space of Nine Years, drawing Forth, and Selecting the true Root, and Marrow of Nature (by a series of Experiences) from its most inward Recesses, and secret Habitations.

II. The which I am moved and compelled to from the singular Good-will, entire Affections, and Sincere Love, which as well as in times past, I now at present bear unto you: And therefore shall faithfully (tho' briefly) declare the Secrets of this Art to you; plainly and openly, not Darkly and Aenigmatically, as if it was done under a Shadow or Vail.

III. Such indeed is your Life (Your Works witnessing the same) that you are as a healing Balm, a Refuge of of Defence, and Shelter to the Church of God, a Pillar of his Holy Temple, for which Reasons, I am obliged to reveal these hidden Mysteries, and make known to you the abscondite Paths of Nature, not to rejoyce your outward Man only, by adding Health and long Life, leaping up Treasures, and external Honours and Applause, in the World, but to excite in you the highest Devotion to God Almighty, that you might become good to all Men, profitable to the Church, a Father to the Fatherless, and a Sanctuary to the Needy and Distressed.

IV. And in these things, I am confident of you, in whom is found such a Portion and Treasure of Vertue, Prudence, Piety, and true Wisdom, but most chiefly, for that I know you to be such a one, who has God always before your Eyes.

V. And therefore I speak truly and fervently, and I will declare the Truth to you, with all faithfulness according to the reality of my Soul; I shall Elucidate the undoubted Verity, and declare such things, as with much Labour, Care, and Diligence I have sought out, and obtained the knowledge of; which I have seen with my Eyes, and have handled with my Hands, and which my own self has done: And in this matter I will neither be tedious nor obscure, lest that love which I profess to you, should seem to be deficient or imperfect.

VI. Whatever I write, I shall open the same briefly and plainly beseeching God, that the matter whereof I shall entreat, may become profitable unto you; and that if you shall please to put the same into practice you may find the faithful experience thereof, and not be deceived, or spend your time in vain: For we know certainly, that of all transitory things, Time is truly the most pretious.

VII.   Wherefore I write unto you (honourable and dearly beloved Friend) such things only as may be profitable; making this humble suit unto your Excellency, that the Revealed Secrets and Experiments which I send you in this little Book, may not be prostituted, or bestowed upon unworthy Men, who are naughty, or swoln up with Pride, or whose Souls are bound up in their Covetousness.

VIII.   I require not of you for this Secret, a great Summ of Gold or Silver; nor do I put this Secret in writing, for you to bestow much Cost and Expences upon it; nor do I for my self desire any reward, these things agree not with the Philosophick Verity, which professes, that its Works are not chargeable and Expensive. *Morienus* saith, beware that you spend nothing in this Magistery of Gold. And *Dastine,* saith with the Value of one Noble is the whole Magistery performed.

IX.   Since then it is so, in what thing is our Gold to be found? Is it not in Mercury, which is called Quick or living Gold? *Raymandus* saith, He that will reduce

Quick Gold into thin water, must make it, doe it, and Work it by its contrary. For saith he, Quick or living Gold, has in its self, four Natures, and four humours or Elements. And therefore saith he, if you putrefie its Cold with its Hot, and its Dry with its Moist, you shall not only have the Humidity of all Bodies, but you shall have a Menstruum, which will dissolve Argent Vive forever. For the least part of Mercury being once dissolved, the dissolved Mercury will always dissolve Mercury ad Infinitum.

X.   (Mercury may as well be called Quick-Gold, as QuickSilver, for it contains them both. If Air will make this Separation, we must put thereto divers contrary things, as Roger Bacon saith in Speculo. But this putrefaction cannot be done, till it is dissolved in Water white as Milk, putrifie that Milk 15 days in B.M. then separate its Element, and cleanse its Earth, and after that joyn it again in equal weight, then is the Elixir made complete for Saturn and Jupiter. Quick-Gold is Crude, Imperfect, and unfixt in every degree, and yet it is accounted a Body, altho' there be no fixation in it, and therefore it may be much sooner brought to its first matter, than any other of the Bodies, that have any part of fixation in them, for they must have much Labour and long time to separate them, and bring them back into their fixt matter.)

XI. For saith *Lully,* The Elements of Mercury may be dissolved, and being so dissolved, they may be separated. There be some that think our Resoluble Seed, or dissolved *Menstrum,* is the water of *Argent Vive,* made only by itself, because it does dissolve both Metals, and pretious Stones which we call Pearls; and so it is. Now how this dissolving Menstrum is made, not only Raymund seems to shew, but *Roger Bacon* in like manner in his Speculum *Alchymia,* where he saith, put the Body which is most weighty, into a Distillatory, and draw forth thereof, its Sweet Ros, or Dew, with a little Wind, or Breath: (for betwixt every drop of Water, comes forth a Breath, as it were of a Man, which is the substance of *Argent Vive,* and which the Philosophers call our Mercury; which if it be well putrefied before hand, will then yield the more, and issue out forcibly, as if it were Wild Fire out of a Trunk, especially when the Red Fume comes.) Thus have you one of our *Argent Vives.*

XII. To the same thing Raymundus assents, where he saith, then have you that Argent Vive, which is called Ours; and so it is indeed one of Our *Argent Vive;* altho' the intent of the same Philosopher in *Libro Animae Artis Transmutatoriae,* Cap. 2 was touching another more noble and more excellent Water (supposed by some, to be Our Burning-Water, drawn out of the Gum of Vitriol,) by the Virtue of which most Noble and

Excellent, attractive Water, he did not only often dissolve the Body of Sol (not as he doth it with the aforesaid *Argent Vive* commonly dissolved) but also the same solar Body, by force of that attractive Virtue, is disposed in a more noble manner; as I myself have seen done, not only in the Metalline Elixir, but also in the Elixir of Life, as here-after shall be declared, Chap. 71, 72. Sect. XIII. It is fancied by an Experienced Philosopher, that Mercury did speak, and said, I am the Father of Enchantments, Brother to the Sun, and Sister to the Moon, I am the Water of Life drawn out of Wine, (i.e. *out of the Wine of Mercury*) I kill that which was alive, and make alive that which was dead; I make Black, and I make White, and I carry in my Belly the *Sol* of Philosophers; and therefore he that can joyn me after that I am dissolved, and made the pure clear and Silver like Water, called *Lac Virginis,* with my Brother the Sun, he shall tinge him with my Soul, not only much more than he was before by an hundred fold, but also if he be joyned with my Sister *Luna,* he shall make all things fair and bright. (this *Lac Virginis* is a Silver-like Water somewhat thick.)

CHAPTER LXII

A FARTHER DISCOURSE OF THE PHILOSOPHERS MERCURY.

I.   Of this Mercury, speaks another Philosopher thus, when its Elements are separated, and again joyned and mixed together by equal weight, then is it made a compleat Elixir upon *Saturn* and *Jupiter;* but its Elements cannot be separated, until such time as it is dissolved: and of this Metalline Water, ought the Artist to draw the *Tincture.*

II.   (The Elements of *Mercury* being separated, and again commixed by equal weight or proportion, make the Elixir compleat with often dissolving and congealing of the Spirit, which must be done upon a Marble Stone, weighing the Body, and then taking its weight of the Secrets Salt, grinding them together very subtil, then putting them into Balneo, that they may be dissolved; which done, take it out, and make your congelation a dry Fire, do so oftentimes, and then, etc.)

III. And therefore to confirm this, *Raymundus* saith, O my Son, Our Tincture is drawn out of one Vile thing, and is decked, finished, and ended with another thing which is more Noble; for we do Ferment it with Vulgar

Gold: He calls it Vile, because he saith it is sometimes found in Vile Places, as in Old draughts: also it is Vile, because (as Raymundus saith) it is found not only in a filthy form, and ugly shape, but because it is in everything, of the which (saith Albertus) is made a Permanent or fixt Water.

IV. (Here is to be Noted, that Raymundus commands this Tincture to be drawn out of the Body of Venus, which Tincture he does Ferment with the prepared *Calx* of common or Vulgar Gold.)

V. And therefore saith *Avicen,* it behoves you to have a great quantity of our Gold, and of our Silver, to the end, that thereby the humours may be drawn forth: viz. to have at the least sixty pounds weight, which will be a sufficient quantity for your whole life. He also saith, the best *Mercury* is brought in skins, from *Mount Pasullane.* Of this *Mercury,* Geber saith, you must labour in all your work to separate *Mercury* or as others read it, to con-vince or over-come *Mercury,* in commixing and conjoyning; for he that cannot destroy *Mercury,* or undoe it in its composure, cannot repair or restore it: nor may you work with it as Raymund saith, till it is dissolved.

VI. And therefore it is said, joyn not that which is Crude, with that which is Decocted; for of that only with the Ferment, is made the Elixir, which does congeal all manner of Argent Vive. Wherefore as Raymund saith, it is never congealed without a congealing *Sulphur;* and being congealed, you have a great secret: for in the dissolved, Decocted *Mercury,* is a great and hidden Mystery.

VII. Another Philosopher also saith, that there is a certain subtil Fume, which does spring forth from its proper Veins, dispersing and spreading its self abroad the which thin Fume if it be wisely gathered together again, and sprinkled upon its proper Veins or Matrix, it will make not only a certain fixation (of which thin Fume, in short space is made the true Elixir) but also cleanses the Impure Metals or A*lchymick* Body.

VIII. (as to the Tincture mentioned at Sect. 4 above, it rather seems by other works of the said Raymundus, that he drew it out of *Quick-Silver*, and no other vile thing, of which *Mercury* is made. What is meant at Sect. 6. By not joyning the Crude with the Decocted, is to be under-stood of not joyning Crude *Mercury* to the Decocted Bodies or Metals, but to put to them Decocted, i.e. dissolved *Mercury*. And herein is hidden a great secret, for *Mercury* being dissolved, is an hot and moist Sperm; but Crude, it is cold and dry *Saturn*. So that if you putrifie its hot and moist

Sperm with its cold and dry Earth, you will have Quick-Silver dissolved, which is not Crude, but Decocted Mercury. So that in Crude Mercury dissolved is hidden a great Mystery. And however it is dissolved by a Fire not natural or against Nature, yet it must be mixed, conjoyned, fixed.)

IX.   This Alchymick Body is called *Leprous Gold,* wherein Gold and Silver, are in Essence and Power, but not in fight or appearance; in its Profundity or Depth, it is Airous or Spiritual Gold, which none can obtain, unless the same Body be first made clean and pure. The which impure Body after mundification, is a thousand times better than are the Bodies of common *Sol* and *Luna,* Decocted by natural heat.

X.   (This Leprous Gold the Philosophers call, *Adrop,* or *Adrup,* which Gold is the Philosophers Lead. This Alchymick Body (in his Concord) he calls *Venus* in the lesser Work, both for Gold and Silver, because it is a Neutral Body, and very easie to be changed to either: and by this the sense of Sect. 4 and 8. aforegoing may be more easily understood. The Earth, the uncleansed Body, is to be putrified with its own Water, and afterwards nourished with its Mothers Milk, which is called the Sulphur of
Nature.)

XI. The first Matter of this unclean Alchymical Body is a *Viscous* Water, which is thickened in the Bowels of the Earth. And therefore of this Impure Body (as *Vincent* saith) is made the great Elixir of the Red and White, whose name is *Adrop,* or *Adrup,* viz. the Philosophers *Lead.* From the which Raymundus commands an Oyl to be drawn: from the Lead of the Philosophers (saith he) let there be an Oyl drawn of a Golden Colour; if you can separate this Oyl (wherein is Our second Tincture and Fire of Nature) from its Flegm, which is it watrishness, and wisely search out the Secret thereof, you may in the space of thirty days perform the Work of the Philosophers Stone.

XII. This Oyl does not only make the Medicine penetrable, being amicable and conjoynable to all Bodies or Corporeal things, but it is also the hidden or Secret fire of Nature; which does so augment the Excellencies of those Bodies to whom it is so joyned, that it makes them to exceed in infinite proportions of goodness and purity. So much as does appertain to the Work of *Alchymiae,* which is only for the Elixir of Metals is now sufficiently opened, which if you rightly understand, you will find that no great cost is required to the performance of this Philosophick Operation.

XIII.   (The Innatural Fire is Our *Aqua Foetens*, or SeaWater, sharp, peircing, and burning all Bodies more fiercely than Elemental Fire, making of the Body of *Sol*, a meer Spirit, which common Elemental Fire has not power to do.)

XIV.   But this Elixir of Metals is not all that I intend to shew you; the Elixir of Life is that which I chiefly designed, infinitely exceeding all the Riches of this World, and to which the most excellent of all the Earthly things cannot be compared. And therefore, I shall 1. Shew in the Mineral Kingdom, the Elixir of Metals, and that after divers manners. 2. In the Vegetable Kingdom, the Elixir both of Metals, and of Life. 3. In the Animal Kingdom, the Elixir of Life only; albeit the same Elixir of Life is most excellent for the transmutation of Metalls.

XV.   There are three things necessary to this Art, of which you ought not to ignorant, viz. 1. The Fire wherewith: (*The fire of Nature, Innatural, Elemental, and which is against Nature,* destroying the special form of all that is dissolved therein.) 2. The Water whereby: (*as in the Compound Water.*) 3. And the thing whereof: (is *made the congealed Earth, as White as Snow*[3]) Of all which in their proper order.

---

[3] *Compare with* Golden Chain of Homer-*HWN*

CHAPTER LXIII
OF THE MINERAL STONE, AND PHILOSOPHICK FIRES.

I. On a time as I have learned, there was an Assembly of Philosophers, where the *Matter* of the Secret *Stone,* and the *Manner* of working it, was propounded. Several spoke their Opinions, but at length, one younger in Years, and (as was thought) Inferiour in Learning, declared his thoughts and knowledge concerning that Secret. I know saith he, the Regiments of the Fires. When they had heard what he could say, they all as amazed held their
peace for a while.

II. At length, one of the Company made answer; If this be true which thou hast said, thou art Master of us all, and thereupon with one consent, they gave him the Right Hand of Fellowship. Whereupon they gathered, that the Secret of this wonderful Tincture lay chiefly in the Fire.

III. But the Fire differs after several manners; one Natural, another innatural or preternatural, another Elemental, another against Nature. The Natural Fire does come from the Influence of *Sol,* and Luna, and the Asterisms, or the *Sun, Moon* and *Stars*, of the which are Ingendred, not only the burning Waters, and potential Vapours of Minerals, but also the Natural Virtues of living things.

IV.   The Innatural or Preternatural Fire, is a thing accidental, as Heat in an Ague, being made Artificially, and called by the Philosophers a moist Fire, Our generating Water, the fire of the first Degree; and for the temperature of its Heat is called a Bath, a Stew, a Dunghil, in which Dunghil is made the putrefaction of our Stone. See
Sect. 13 of the former Chapter, where it is more amply defined.

V.   The Elemental fire, is that which does Fix, Calcine and Burn, and is nourished by Combustible things.

VI.   The fire against Nature (which is a violent strong, Corrosive, destroying the special form of that which is dissolved therein,) is that which in Power Dissolves, Frets, Infects, and destroys the generative Power of the form of the Stone: it does Dissolve the Stone into Water of the Cloud, with the loss of its Natural, Attractive, and special Form, and is called *Fire against Nature* (as Raymundus saith) from its Operation: for that which Nature does make, this fire against Nature destroys and brings to Corruption, unless there be fire of Nature put to it.

VII.   Here as Raymundus saith, lies contrary Operations, *(as in the Compounded Water)* for as the *fire against Nature,* does Dissolve the Spirit of the fixed Body; the Volatile Spirit is thereby constrained to retire into a fixed Earth, (a *Congealed Earth as White as Snow.*)

VIII.   For the fire of Nature does Congeal the Dissolved Spirit of the fixed Body into a glorious Earth: and the Body of the Volatile being fixed, by the same *fire against Nature,* is here again by *the fire of Nature* resolved into the Water of Philosophers, but not into the Water of the Cloud: and so by this means the fixed is returned back again into its wonted Nature of Flying, and the moist is made dry, and the ponderous is made light.

IX.   But yet he saith, *this fire which is against Nature* is not the Work of Our Magistery, but it is *the fire which is purely Natural.* This he saith, because he would shew us thereby the difference between the Mineral Elixir, and the Vegetable, and the Animal. For that these three several Elixirs are made of three several Waters, viz. Mineral, Vegetable, and Animal, which serve for the Work divers ways.

X.   And First we will Treat of the Mineral Elixir, then of the other in order. The Fire against Nature is a Mineral Water, (viz. the Humour or Tincture drawn out of Body of Venus Dissolved in its Mineral Spirit) very strong and Mortal, serving only to the Mineral Elixir.

XI.   This Mineral Water, or Fire against Nature, is drawn with fire Elemental, from a certain stinking *Menstruum,* as Raymundas saith, and is made of four things. It is the strongest Water in the World, whose only Spirit, (saith he) does wonderfully increase and multiply the Tincture of the Ferment: for here Sol or Gold is Tinged with the Mineral Spirit, the which Mineral Spirit is the strength of the most simple *Sulphur* without much Earthiness.

XII.   (Thin Mineral Water is the dropping of *Adrop* or *Adrup, Venus,* which is the noble Tincture called the natural Roman Vitriol, and which for the abundance of its noble Tincture, is called *Roman Gold.)*

XIII.   This some do call the Spirit of the *Green Lyon,* others the blood of the *Green Lyon:* wherein almost all Err, and are deceived: for *the Green Lyon of the Philosophers,* is *that Lyon,* by whose Virtue attractive, all things are lifted up from the Bowels of the Earth, and the Winter-like Caverns, making them to Wax green and flourish: whose Child (for all the Elixirs are to be had from it) is to us most acceptable and sufficient.

XIV. (The Child of Philosophers is generated of their Green Lyon, of which Child is had the strength of Sulphur, both White and Red; Our two Sulphurs of Nature are the Gold and Silver of the Philosophers, and their hidden Treasure.)

XV. Of this Child of the Green Lyon of the Philosophers is drawn the strength of Sulphur White and Red, but not Burning as Avicen saith, which are the two best things the Alchymist can take to make his Gold and Silver of: and this is sufficient to be said, for the attaining the knowledge of the Green Lyon: which is so called, because, that when he is dissolved, he is streight ways adorned with a green Vesture. (i.e. *When our* Sulphur *of Nature is dissolved in its own* Menstruum, *which is* the Virgins Milk, *it is clothed with this greenness, and therefore called* the Green Lyon.)

XVI. But of the *Green Lyon of Fools,* this we say, that from it with a strong, fire is drawn *Aquafortis,* in the which, the aforesaid Philosophers Lyon of the Mineral Stone, ought to be Elixirated, and assumes its Name. Raymundus saith, it were better, or safer, to eat, the Eyes of a *Basalisk,* than that Gold, which is made with the Fire against Nature.

XVII.   And I say also, that the things from whence the same *Aquafortis* is drawn is green Vitriol and *Azoth:* i.e. Vitriol Natural, not Artificial, viz. the droppings of Copper, *called also Roman Vitriol, Roman Gold,* by many of the Philosophers, from the abundance of its noble Tincture, the which Tincture must be Fermented with common Gold.

XVIII.   How great and Secret a Virtue, then, and of what strength, the Fire against Nature is, evidently appears in the construction of the Body of the Volatile Spirit, being by its vulgarly sublimed in the form of Snowy Whiteness. Raymundus in the end of the Epistle of his Abridgment saith, feed Argent Vive with this Oyl, viz. with the Oyl, wherewith the Spirit of the Quintessence is thickened, etc.

XIX.   For want of such, Natural Virtiol, the true and natural Principle, not Artificial, (as Vincent saith) made of Salts, Sulphurs, and Alums, which cut and gnaw Metals, is to be chosen, lest in the end of your work you fail of your desire. (The Philosophers will you to Calcine Sol with Mercury Crude, till it be brought into a Calx Red as Blood: Here comes in the work of Sol and Mercury together, brought into a dry Red Pouder and fixed, but whether it is to be done with Mercury or Sulphur, the Water of him, is doubtful.)

CHAPTER LXIV

THE MANNER OF

ELIXIRATION WITH THE FIRE AGAINST NATURE,

I. Take the first Sol, Calcined with the first Water, viz. the Mercurial Spirit, very clean, and brought into the Color of Blood, in the space of 20 days, (in lesser time it is not to be done.) This Calcination cannot be so profitable, as it would be, unless Sol be first Mercurializ'd into such a thinness, as it may cleave together to that to which it must be joyned in a 24 fold proportion, (viz. as 1. to 24.) strained through a clean Linnen Cloth, without any remaining substance of the Gold.

II. I myself have seen it so ordered and done; and then it may certainly, in a strong Bolt-Head, well Luted on every side, except on the Top, boyling in a strong Fire for the space of 20 days, be precipitated into a Red Pouder, like Cinnaber, (all which I have seen performed.) Every particle of this Pouder you shall so fix, as that if it be put upon a Red-Hot Iron Plate, its Spirit shall not fume or fly away.

III. This Pouder Dissolve with, or in our Fire against Nature; being Dissolved, abstract the Water of the Fire against Nature from it, so long till the substance of the Pouder so Dissolved, do remain in the Vessel, as thick as an Oyl; which Oyl, first, with a soft fire, and after with a stronger, fix into dry Pouder.

IV.   (This Work is not to be done all at once, but by little and little at a time, till it goes through with it in the Color of Blood; then will it precipitate into a Red Pouder, called by the Philosophers Sericon: Dissolve it with as much of Our Vegetable Sal Anatron, the space of an hour, then set it in balneo, in a long Receptory, till it be cleanly dissolved, and becomes as it were a fine Wine, which with the very softest heat, make it to Evaporate, and Congeal, so will you have a pure Stone, and of subtil parts.

V. Also if you dissolve this same Red Pouder of Mercury in Water or Spirit of Common Salt, prepared as Bachon and Albertus have taught, you shall have an Oyl or Salt of Gold, which no Fire can destroy, which will melt and tinge with a solar Color upon a Plate of Venus. This Treasure carry always with you, wheresoever you go: Who knows not the Secret of this prepared Salt in Our lesser Works, knows little of the hidden things of Alchymie.)

VI. Try this fixt Pouder (at Sect. 3. above) for the

fixation, reiterate still the Work with the same Fire against Nature upon the same Pouder Ten times, and it will be dryed up no more into Pouder, but remain in a thick Oyl, the which will turn Argent Vive, and all Bodies into pure Alchymick Gold, sufficiently good for all works of the *Goldsmith,* but not for Medicine for Man's Body.

VII. A Second way, Gold is much more wonderfully Elixirated by the said *Fire against Nature,* compounded with the *Fire Natural,* after this manner. Let Vitriol of the Fire of Nature, made of the most sharp Humidity, or moisture of *Grapes,* and *Sericon,* joyned together in a Mass, with the Natural Mineral Vitriol (*called the Gum of* Adrop, or *Vitriol* Azoth,) made somewhat dry, and with Sal Nitre, be dissolved.

VIII. First Ascends a Fair, Weak, Flegmatick Water, which cast away. Then a White Fume, making the Vessel appear White like Milk, which Fume must be gathered into the receiver, so long till it ceases, and the Vessel becomes clear, of its own Color. This water of the White Fume is the stinking Menstruum, which is called Our Dragon against Nature. This Menstruum, if the said Dragon against Nature was absent, would be our Fire Natural, of which we shall hereafter speak in its proper place.

IX.   (Raymundus saith, *this Water is made of four things*: 1. *The Composition of* Sal Amarum. 2. Menstruum Foetens. 3. Argent Vive, *which is a common substance in every combustible body.* 4. Mineral Vitriol.

X.   This compounded Water Mineral, and Water Vegetable, being mixed together, and made one Water as aforesaid, doth work contrary Operation, which is wonderful, it Dissolves and Congeals, it makes moist and dry, it putri-fies and purifies; it divides asunder and joyns together; it destroys and restores; it kills and makes alive; it wounds and heals again; it makes soft and hardens; it makes thin and thick; it resolves Compounds, and Compounds again: It begins the Work and makes an end of the same.

XI.   These two Mineral Waters Compounded together in one, are the two Dragons Fighting and striving to gather one against the other in the Flood of *Satalia:* viz. The White Fume and the Red; and one of them shall devour the other. And here the Solutory Vessels ought to be Luted but gently, or closed with Linnen Cloth, or with Mastick, or common Wax, or Cerecloth.

XII.   These two Dragons are Fire and Water, within

the Vessel and not without; and therefore if they feel any exteriour Fire, they will rise up to the top of the Vessel, and if they be yet forced by the violence or strength of the Fire, they will break the Vessel, and so you will lose all your Work.

XIII.   This Compounded Water aforesaid, does Congeal as much as it does Dissolve, and lifts it up into a glorious Crystalline Earth. This is our Secret dissolution of the Stone, which is always done with the Congelation of its Water. The Fire of Nature is here put to the Fire against Nature; therefore as much as the Stone has lost of its form by the power and strength of the Water, or Fire against Nature; so much has it gotten and recovered again of its form, by the Virtue of the Water, or Fire of Nature. But the Fire against Nature, by the means of the Fire of Nature, cannot be destroyed.

CHAPTER LXV

THE PRACTICE WITH THE SAID COMPOUNDED WATER, UPON THE CALX OF THE BODY DISSOLVED.

I. The Practice with the said Compounded Water, upon the Calx of the Body duly dissolved and prepared: Take the prepared Body (made with a thick Oyl,) put to it so much of the Compounded Water as may cover the same Calx (i.e. Our prepared Calx with Our Vegetable Menstruum) by the depth of half an Inch. The Water will presently boil over the *Calx* without external dissolving the Stone, and lifting it up into the form of *Ice,* with the drying up also of the said Water.

II. The said *Calx* being so dissolved and sublimed into the form of *Ice*, you must take away; after this is done, the residue of the Calx remaining in the Vessel, undissolved, shall again be well dryed by the Fire, upon which put so much of the said Compounded Water as you did before, dissolving, subliming and drying, till the Calx is wholly dissolved.

III. The substance thus dissolved, subtily separated and brought into a Pouder, must be put (as thereafter shall be shewed) into a good quantity of the Fire of

Nature (*which is a Quintessence*) the same being first well rectified, and the Vessel well stopp'd, to the end, that the means of the heat outwardly administred unto it, pro-curing the inward heat to work, it may be dissolved into an Oyl; the which will soon be done, by reason of the simplicity of the Water, or simple Fire of Nature.

IV. And therefore when you have brought the said Pouder so dissolved, sublimed, and prepared with the said Compounded Water into an Oyl (then is our Menstruum Visible unto sight) by putting thereto a good quantity of the aforesaid rectified simple Fire of Nature, as before declared; then abstract or draw away the said Water again from the same Oyl, by Distilling the same in a moist, Temperate heat, so long till there remains in the bottom of the Glass a thin Oyl.

V. This Oyl, the oftner it is dissolved with the said simple rectified Fire of Nature, and the said Water Abstracted or Distilled by a Temperate heat, so much the more will the said Oyl be made subtil and thin.

VI. With the said Oyl (provided the Calx be the Calx of Sol or Luna) you may incere the substances or Calces of other Bodies, the said Bodies being first

dissolved, exalted, sublimed, and prepared with the said Compounded Water, in manner and form of Ice aforesaid, till that by the Inceration of the said subtil and thin Oyl of Sol and Luna, the said substances of other Bodies be made fixed,
and to flow like Wax.

VII. With which flowing substance, you shall not only congeal Argent Vive into perfect Sol and Luna, according as you have prepared your Medicine, but you shall also with the same fluxible and flowing substance, transmute and change all such other imperfect Bodies, (as they were, whose Calces were so sublimed, and from whom, at their first sublimating or lifting up, they took their beginning)
into Sol and Luna aforesaid.

VIII. And this thin and subtil Oyl, being put into *Kemia* its proper Vessel, first sealed up, to putrifie in the Fire of the first degree, being moist: it becomes as black as liquid Pitch. The fire may then have its Action in the Body, to corrupt it, (the same Body as before so opened.)

IX. Therefore it grows first black, like melted Pitch, because the heat working in this moist Body, does first beget a blackness, which blackness is the first sign of Corruption: and since the Corruption of one thing is the generation of another; therefore of the Body corrupted, is generated a Body Neutral, which is certainly apt, declinable, and applicable unto every *Ferment* whatsoever you please to apply it to.

X.   But the *Ferment* must be altered together with the *Alchymick* Body; and the whole substance of our Stone or Elixir must partake of the Nature of the Quintescence, otherwise it will be of no effect.

XI.   And between the said sign of blackness and perfect whiteness, which will follow the said blackness, there will appear a green Color, and as many variable Colors afterwards as the mind of Man is able to conceive.

XII.   When the present White Color shall begin to appear like the eyes of fishes, then may you know that Summer is near at hand, after which *Autumn* or *Harvest* will happily follow with ripe fruit, which in the long looked for Redness: This is after the Pale, Ashy, and *Citrine* Color.

XIII.   First the Sun does perfectly Descend by its due Course, from its Meridional height and Glory, through its gross and natural solution into an imperfect Pale, and Ashy Color, shining in the Occidental parts of the West, which is somewhat of a yellowish or Brick dust Color: from thence it goes to the Septentrional parts of the Earth, being of a Variable watrish blackness, of a dark, cloudy, alterable, putrefaction watrishness.

XIV. Then it Ascends up to the Oriental parts, shining with a more perfect Crystalline, Summerlike, and Paradisical White: Lastly, he Ascends his *Fiery Chariot,* directing his Course up again to his Meridional Life, Perfection and Glory, there to Rule and Shine, in fire, brightness, splendor, and the highest perfection, even in the highest, most pure, and Imperial Redness.

XV. When this aforesaid simple Oyl of the altered Body, being in its Vessel duly sealed, is by the Fire thus disposed, what is there more than one simple thing, which nature has made to be generated of *Sulphur* and *Mercury* in the Bowels of the Earth?

XVI. Thus it is evident, that our Stone is nothing else but *Sol* and *Luna, Sulphur* and *Mercury:* *Male* and *Female; Heat* and *Cold.* And therefore (to be more short) when all the parts of our Stone, are thus gathered together, it appears plainly enough, what is our *Mercury*, Our *Sulphur,* Our *Alchemick* Body, Our *Ferment,* Our *Menstruum*, Our *Green Lyon:* And what Our *White Fume,* Our two *Dragons,* Our *Fire,* and Our *Egg,* in which is both the Whiteness and the Redness.

XVII.   As also what is *Man's Blood,* Our *Aquae Vitae,* Our *Burning Water*, and what are many other things, which in this Our Art are Metaphorically, or figuratively named to deceive the Foolish and unwary.

XVIII.   Also there is a similitude of a *Triune,* shining, in the Body, Soul and Spirit. The Body is the substance of the Stone. The Soul is the *Ferment* which cannot be had, but from the most perfect Body; and the Spirit is that which raiseth up the Natures from Death and Corruption to Life, Perfection and Glory.

XIX.   In *Sulphur,* there is an Earthiness for the Body: In Mercury, there is an Aerealness for the Spirit, and in them both a Natural Unctuosity for the Soul or Ferment; all which are inseparably United in their least parts forever: From this Fermental Body the Stone is formed, and without it, it cannot be made.

XX.   It is the peculiar property of *Sol* and *Luna,* (the which property appertains to the Stone itself) to give the form of Gold and Silver. And therefore the Elixir, whether it be White or Red, may be Infinitely augmented with the Fermental Oyl: if you do cast the same upon Mercury, it shall transmute it into the Elixir, which Elixir must be cast afterwards upon the Imperfect Bodies.

XXI.   Moreover the said White Elixir is augmented with Mercurial Water, and the Red Elixir with the Mercurial Oyl; the which two, viz. the Mercurial Water and Mercurial Oyl, can only be had of Mercury dissolved of itself.

XXII.   See what the Scripture saith, *He stroke the Stone and Water flowed out, and he brought forth Oyl out of the Flinty Rock.* We may Note the whole composition of the Elixir in these four Verses following.

XXIII.   *He stretched forth the Heavens as a Curtain. The Waters stood above the Mountain:* This is the Water which does cover Our Matter, and performs the dissolution thereof, causing a cloudy Ascension. *That does walk upon the Wings of the Wind.* This figures forth the sublimation of our Stone.

XXIV.   *Who makes his Angels Spirits, and his Ministers a flame of Fire.* By this is shadowed forth the rectification, separation, and disposition of the Elements, *who has founded the Earth upon its Basis; so fixt, that it shall not be moved forever.* Under which is described the fixation of the Elements, and the perfection of the Philosophick Stone.

CHAPTER LXVI

ANOTHER WAY OF ELIXIRATING GOLD BY THE FIRE AGAINST NATURE.

I.   Another way, by which the Body of Gold is Elixirated by the power of the *Fire against Nature,* through the help of the Operation of the *Fire of Nature;* which is thus. Dissolve the Body of pure Gold in the *Fire against Nature.* the same fire being well rectified *Arsenick* (<u>Mercury sublimate</u>) as the manner is; from which Gold being so dissolved into a Citrine, clear and shining Water, with-out any Heterogenity or Sand remaining, let the water be abstracted, till the Body does remain in the bottom of the Glass, like a fixt Oyl.

II.   Upon this Oyl, affuse the said *Water,* or *Fire against Nature* again, and abstract again, and this work so often repeat till the *Water* or *fire against Nature,* have no more sharpness upon the Tongue than common Well-Water.

III.   This done, draw such another *new Water* or *fire against Nature,* which Affuse upon the former Oyl, and abstract in all respects as before is taught: then Affuse upon this Oyl the *Water* or *fire of Nature* well rectified, and let it be double in quantity or proportion of the said Oyl of the Body so dissolved, and put it into a Vessel, which stop well, and set it in Balneo for seven days; so will the *Water* or *fire of Nature* become a Citrine Color.

IV. This *Water* or *fire of Nature* by its attractive Virtue, will draw away the Tincture from the fire against Nature, as Raymundus saith in his practical Alphabet. And altho' it is somewhat opposite to Nature, to dissolve the Bodies with the *fire against Nature;* yet if you know how to comfort the matter with the *fire of Nature,* and by Balneation in 15 days, to draw it from the blackness of the water, or *fire against Nature,* (the which may be done, as I have proved, in 6 days) you shall perfect the work, and attain the desired end.

V. Let the aforesaid Natural Water or fire of Nature, so tinged with a Yellow Color, be always warily emptied, and poured off from the aforesaid dissolved Bodies, into another Vessel, with a narrow Mouth, that may be firmly stopped: and then with more of the said fire, let there be made in Balneo, in the space of time aforesaid, another quantity of the said Oyl.

VI. And so the same water being tinged with Sol or Gold, let it be warily emptied, and poured off as before: and when the water of Nature will be tinged no more, then it is a sign, that the Tincture is all drawn forth from the dissolved Body by the Fire against Nature.

VII. Put the Tinctures thus decanted off into a Glass Stillatory, and with a soft or easie Fire abstract the Water or Fire of Nature from the same, so long till you see in the bottom an Oyl; to which you must put New Fire of Nature again, well rectified: and after the Matter has stood in Balneo for the space of 6 days, then abstract the said water or fire of Nature by distillation.

VIII. And let the work with the same water be repeated upon and from the said Oyl, after the same manner so long till you have brought your Oyl of Gold to be most subtil and pure, without any Foeculent grossness, wherein let nothing of the water or fire of Nature be left behind, but the substance of Gold only, turned to Oyl.

IX. This subtil and pure Oyl of Gold, being put in *Kemia,* or its proper Vessel, and firmly sealed up, may be the aforesaid Regiments be changed into the great Elixir, as it is shewed before with the other simple Oyl, made with the Compounded Water, in the former practice, at Sect. 8. Chapter 65. aforegoing.

X. But to proceed: sublime *Quick-Silver* with *Roman* Vitriol and prepared or Calcined Salt; and after that sublime it by itself alone three times from its Foeculent substance. This done, and the same made into Pouder, put this sublimate Pouder into a fixatory Vessel, and put thereto a certain quantity of your aforesaid Oyl of Gold, but so much only, as may scarcely cover the sublimate: firmly close the Vessel, and set it in a soft Fire, till the Natures are perfectly joyned together.

XI. This done, grind it upon a Marble, and Incinerate it again with your said Oyl of Gold, and after put it again into its Fixatory Vessel, under a fire of the first Degree as before, and let the same Vessel stand twice as long as it did before, to the Intent that the Natures may be firmly Compact and United together.

XII. Now this Rule is generally to be Observed: that the Vessel with the Matter in it to be fixed, ought always to be set over the fire from time to time to be augmented, and this Incineration to be continued still upon the *Argent Vive* sublimed, until the same is perfectly fixed with the said Oyl or substance of Gold.

XIII. The which must be proved upon a Plate of Silver Red Hot: And if it be found fixed, let it have for the greater certainty, one Incineration more of the said

Oyl, which set under a strong fire for the space of three days: then grind it with your Oyl upon the same Stone, till it be as thick as an Oyntment; which make perfectly dry with an easie fire, and then let it be Calcined with a strong fire for the space of eight hours.

XIV. Which done, then Incinerate it, and dry it again with a soft or gentle fire oftentimes, till it stands in the fire like melted Wax. This medicine will transmute Silver substantially and perfectly into fine and pure *Alchymick Gold* perfect to all the works of *Goldsmiths* but not to Medicine for Men.

CHAPTER LXVII

TWO OTHER MINERAL ELIXIRS, OR TWO OTHER PROCESSES OF MERCURY.

I.   There be many other Noble and Profitable Secrets in this Art, or Mystery of our Mineral Stone; viz. good Elixirs to be made out of Metalline Bodies; of which Mineral Elixirs, two are more excellent than the rest, the first of which we shall handle in this Chapter.

(Here *comes the Process or Practical Operation of Mercury mentioned* Chap. 61. Sect. 13. *aforegoing.*)

II.   The first of these Elixirs is only in Mercury: The second, in Mercury and the White Body for the White Elixir; and with the same to the Red too, if you so please, being prudently pursued and sought after.

III.   The first manner to Elixirate only with *Mercury* is thus. Dissolve *Mercury* only, by itself into a Milky water, with the which *Mercury* so dissolved, you may dissolve so much more *Mercury,* and so continually, as long as you please.

IV. Put this into a gentle Fire to be Distilled, so shall you have Our Virgins Milk White and Crystalline, wherewith all Bodies may be dissolved into their first Manner, Washed and Purged.

V. This water is of a Silver Colour, which if you fix with its Earthy Faeces Calcin'd, and after that dissolved again in the quantity of its remaining water, and then again Coagulated and Congealed, (which work is to be done upon a Stone) you will have at length the Elixir of *Argent Vive,* which will transmute all Imperfect Bodies to a perfect Whiteness.

VI. And so of this Mercurial substance is made a water permanent or fixt, wherewith the Calces of all Bodies may be so depurated and Whitened, as thereby to become the most pure and fine Silver.

VII. And therefore as I have said before in the beginning of this work, when *Mercury* is dissolved, then are its Elements separable; and after the separation of its Mercurial Liquor, and that a competent putrefaction is performed; after the same White Liquor, there will Distill a Golden moisture or humour, to which if you add a small quantity of the Ferment of the Gum of the aforesaid Elixirated White Stone, that then the same White Stone, with the said Golden humour, shall be made the Red Stone, which shall transmute *Argent Vive,* and all Metalline Bodies into the finest and most pure Gold.

VIII. Again, if you take the aforesaid Red humour of *Mercury* and Dissolve in it a little of the aforesaid Red Ferment, being made as aforesaid of the White Stone, and then with the same Red humour of *Mercury,* so Fermented with itself, the Calces of all Bodies, may be so depurated and Citrinated, that thereby they may become most pure Gold.

IX. When also *Argent Vive* is dissolved, then dissolve in it a little of the aforesaid Red Ferment, and so put all into *Kemia,* or a proper Vessel, which firmly close up with a Philosophick Seal. Then with a continual and easie or gentle Fire, draw out the Chariot of the four Elements through the Depth of the Sea, until (the Floods being dryed up) there appears in the Matter a bright shining substance, like to the Eyes of the Fishes.

X. For by this Operation, if you keep your Temperate Fire continually alive, the Floods shall dry up, with an exceeding drought, and the dry Land or Earth shall appear, as in the days of *Noah,* the waters were dryed up from off the Earth, and behold the Face of the Ground was dry. And by lifting up the Rod of *Moses,* and stretching out his hand, the waters were dryed up, and the dry Ground appeared in the midst of the Sea: for so says David, *He Rebuked the Red Sea, and it was dryed* up; *he led them through the Depths as through the Wilderness.*

XI.   And then by the Space of Forty days following, it shall be Rubified, (as the Philosophers Demonstrate) by the help of a Vehement Fire, as the Nature of it requires, continuing and remaining in the same strong Fire till it melt and flow like Wax, whereby it will be able to transmute all Bodies into pure fine Gold.

XII.   And thus the White and Red Medicines are multiplied with their own proper humidities: viz. only by the solution of the White Medicines in their own proper White and Red humours, and by their Coagulation again of the same, as necessity requires. Thus have we explicated with singular plainess of Speech, the Elixiration of Mercury per se, or Argent Vive alone.

CHAPTER LXVIII

THE SECOND OF THE FORMER ELIXIRS, WITH MERCURY AND THE BODY ALCHYMICK.

I. To Elixirate with *Mercury* and the Body *Alchymick*. Take One *part of the most pure Kibrick* (quod est pater Mercurii & omnium Liquabilium,) Sea water twelve parts, in which dissolve the *Kibrick:* being dissolved, strain the water through a Linnen Cloth; and what remains undissolved, which will not go through, put into the Vessel called *Kemia,* set it over a gentle fire, as it were the heat of the Sun, untill there appears on the Top a Red Color.

II. Then put to it a quarter more of the Seawater aforesaid, being kept in a very clean Vessel, set it on a very gentle fire, and dry it up again, as you did before, by little and little at a time.

III. For in this Work, by so much less there is put of the Spirit, and more of the Body; for so much the sooner and better shall the Solution be made; the which Solution is made by the Congelation of its water.

IV.   And therefore as the *Rosary* saith, you must beware that the Belly be not made over moist, for if it be, the water shall not receive or attain to its dryness.

V.   This manner of Imbibition must be Observed and continued so long, till the whole water by several Imbibitions shall be dryed up into a Body.

VI.   This done, let the Vessel be firmly and Philosophically Sealed up, and placed in its proper Fornace, with a mean or gentle fire, which must not wax cold, from the first hour you begin to set the same into the Fornace, till you have made an end of the whole work.

VII.   And when the matter is sublimed, then let it be made to Descend by little and little without Violence, the fire being Artificially made or set over it; which done, let it be again sublimed as before.

VIII.   And so let the Soul of the *Sun* of the Vulgar (the which Soul is Our unclean Oyntment, the Spirit not yet conjoyned with the Body) Ascend from the Earth to the Heaven; and again make it to Descend from Heaven to the Earth, till all becomes Earth, which before was Heaven.

IX.   To the end there may be made a substance, not so hard as the Body, nor yet so soft as the Spirit; but holding a mean disposition, standing fixed and Permanent in the fire, like a White piece of melted Wax, flowing in the bottom of the Vessel.

X.   The which White substance, of a mean or middle consistency, must be fed and nourished with Milk and Meat, till the quantity thereof be increased according to your desire.

XI.   This Medicine being Fermented to the Red, with a portion of *Sol* Dissolved in the water of the Sea, by reason of separating the first; the form from the Matter, to the end, that it may be in a more noble form than it was before, when the first qualities did remain undivided; and that it may be brought into a Purple Colour by the help of a strong and continual fire: whereby is made the true Elixir, both for the White and Red Work.

XII.   Now this Elixir, be it White or Red, shall be increased an hundred fold more, both in Virtue, and Goodness, if its Quintessence be fixed with it, and that then afterwards it be brought and reduced by the Fire of Nature into a thin Oyl, the which must be done in a Circulatory Vessel: for truly, then the least drop thereof does Congeal a thousand drops of

Mercury into the very greatest Medicine.

CHAPTER LXIX
OF THE VEGETABLE STONE.

I.   The Vegetable Stone is gotten by Virtue of the Fire of Nature, of the Composition of which fire we now intend plainly to treat and of the way how to work with it, in every respect.

II.   (Its Composition is of four things, as Raymundus saith, in his *Book of Quintessences*: It is a Composition of *Sal Amarum,* which is *Ignis adeptus,* a fire that is gotten without Wood or Coal, and by an easie working, does work against all manner of sharpness of Action of the Visible Fire, like as if it were the fire of Hell; and therefore, altho' Wine be hot, yet this water of Mercury is hotter: for it is able to dissolve all Bodies, to putrefie, and also to divide the Elements, which neither common Fire nor Wine can do.

III.   Some think that this Fire of Nature is extracted or drawn from Wine, according to the common way, and that it must be rectified by often Distillations, until its Flegm is wholly abstracted, which hinders its Heat, Virtue, Strength and Burning. But this, when it is done to all advantages, and its highest perfection (which Fools call the *pure Spirit*) and then put to the *Calx* of the Body never so well prepared, yet will it be weak and ineffectual to Our purpose, for Dissolution, Conservation, etc.

IV. (The true and *Pure Spirit* is Our Silverish Spirit of Wine, which is our Vegetable Mercury, and the true water of the Philosophers. Concerning which, see in Ripley's secret Concord.)

V. Wherefore since the vulgar Spirit or Wine is such, it is evident that there is an Error in choosing of this Principle: for the true Principle, (which is the beginning) is the RESOLUTIVE MENSTRUUM (*which is the Soul of Mercury, and this Tincture is a very Oyl, separate from its foul Earth and faint Water*) which, as we know, and according to the traditions of the Wise Philosophers, is an Unctuous moisture, which is the nearest Matter of Our Vegetable and Philosophick Mercury.

VI. The which Principle, *Resolutive Menstruum, Near Matter,* or *Unctuous Moisture,* Raymundus (in Cap. 6 and Cap. 8. of his Clavis) does call Black, Blacker than Black: The which Black thing or Matter I certainly know.

VII. But since Raymundus saith, that this Resolutive Menstruum, does come from Wine, or the Lees, or

Tartar thereof, how is he to be understood? Truly, he himself unfolds the Mystery: Our Water or Menstruum, is a *Metalline* Water, generated of a Metalline Matter only: So that Raymundus speaks, either of the *Resolutive Menstruum* or of the *Resoluble Menstruum*.

VIII.   (This *Menstruum* springs from a Silver Wine, which does Naturally make a dissolution of its own *Sulphur*. It is apparent in the 11. Cap. of *Raymundus*, that Our Mercurial and Radical moisture is not only Congealed into perfect Metal, by Vapour of its hot and dry *Sulphur,* but that also the same Metalline Water, being so terminated in the form of a Metal, after its Resolution in Ashes has power naturally of a *Menstruum* to dissolve Our *Stone* or *Sulphur,* and change it to its Vegetable Nature, without prejudice or hurt to its own Nature.

IX.   (Wherefore he says, that from whatsoever anything does spring or grow by Nature, that into the same it may again be resolved.)

X.   If he (viz. *Raymundus)* speaks of the first water or Resolutive Menstruum; you are to understand that it is (so as he speaks) not a Metalline Water, but after a certain manner: for this water of the Resolutive Menstruum, is both a Sulphurous and a Mercurial Vapour *(Ignis and Azoth)* and by reason of its Sulphurity, it burns with the fire.

XI.   (This *Resolutive Menstruum* is our Vegetable *Mercury,* which is our Vapourous *Menstruum,* and every burning water of Life, *Aqua Vitae ardens,* by whose attractive Virtue, the Body of the Volatile Spirit, being fixed by the fire against Nature, is dissolved naturally into the water of Philosophers, and exalted and lifted up from its Salt and Combustible Dregs into a clear Mercurial and Natural substance, which must be Fermented with the Oyl of *Sol* and *Luna,* and then is made thereof the great Elixir; with which *Mercurial* substance we also counterfeit Pearls and Pretious Stones.)

XII.   We see also, that in *Tartar* dryed only in the Sun, there are certain Mercurial Qualities shining and giving of light to the Eye, but the kind of Metals is a Composition of *Sulphur* and *Argent Vive.* And therefore, if he means after this sort, then the Resolutive Menstruum, may be taken for a Metalline water; for otherwise it is not Answered.

XIII.   Again, *Raymundus* proves clearly to the contrary, where he answers him who demanded of him; *in what is the Vegetable* Mercury, in *Gold or in Silver? It is,* (saith he) *a simple Coessential substance*, the which is brought from its own Concrete parts and proper Veins, to such a pass or point by the Dissolutive Menstruum, that by Virtue of the simple and Co-essential substance, they are able to multiply their similitudes in Mercuries, which have none in themselves, and are also apt Medicines for Mens Bodies, and to expel and put away from them many Diseases, & to restore to the Old and Aged, their former Youth, and preserve them in Health so long a time as God has designed them to Live.

XIV.   (This Coessential substance is Our White and Red Tincture by whom these Earths that are wanting, are multiplyed in Tincture, whereby they are made Elixirs, to purge Metals, and a Medicine for Man's Body.)

XV.   Therefore, *Our true Metalline Water* is an Unctuous humidity of the Body dissolved to the similitude of Black Pitch, Liquid and Melted; and this Unctuous and Black humidity is called *the true Resoluble Menstruum*. And because we shall afterwards demonstrate the true Resolutive Menstruum, required in this Work, we will here only declare from what principles, and how the said *Resolutive Menstruum* is drawn.

XVI. (Our Metalline Water is separated from the Body of LUNARIA, which is its terminated and Radical humidity in the kind and Color of White shining Silver, and its Body, is Our black *Sulphur:* Therefore see Chap. 63. in the Lunary Branch, and in his Clavis where you will find the Radical humidity to be the true Menstruum wherewith the solemn dissolution of its own black Body is made.)

XVII. Raymundus doth say, that an Unctuous Humidity is the last comfort and support to the Humane Body, which what it is, is manifest to the Philosophers; it makes a noise or sound in the Vessel, and is Distilled with a great deal of Art. He also saith, that Our Stone is made of the hottest Matter or substance in Nature: And I say that Wine is hot; but there is another thing which is much hotter than Wine, whose substance, by reason of its exceeding Airyness or Spirituosity is most quickly inflamed by the Fore.

XVIII. And the Lees, or TARTAR, and Dregs of this Unctuous humidity, is gross, like the Rinde or Bark of a Tree: and the same Tartar is blacker than the Tartar of the black Grape of Catalonia, for which cause it is called by Raymundus, a Black, more Black than Black. *(By these Lees, or Tartar and Dregs, is meant the Lees of our Silver Wine, separated from the Lunary Body).*

XIX. And because that this humidity is Unctuous, therefore it better agrees with the Unctuosity of

Metals, than the Spirit drawn from Common Wine; for through its Liquefactive Virtue, Metals do Melt, and are made flowing and moist in the Fire; the which Operation truly the Spirit of Common Wine cannot do.

XX. For the Spirit of Wine, how strong soever it be, is (comparatively) but clean Flegm or Water: whereas contrariwise, in Our Unctuous Distilled Spirit, there is no watrishness at all. But this thing being rare in our parts, as well as other Countries, *Guido Montanor* found out another Unctuous humidity, which swims upon other Liquors, which humidity proceeds from Wine, which *Raymundus*, and *Arnoldus* knew, with some others, but they taught not how it should be obtained.

XXI. (Our Tincture in Distilling, is separated both from the Flegm and its gross Foeces, till it be like an Oyl, and that is the Soul of Mercury, which is Air and Fire, separate from its two extreams; and so it being an Unctuous moisture, is the mean. See the first and last chapter of Raymund's Codicil.)

XXII. Notwithstanding, Raymundus saith, it must be drawn from Death, and from the Faeces of Wine by rectification, that it may be acuated in Distillation by hot Vegetable substances, thereunto appertaining, as Pepper, *Euphorbium,* etc. for without these things he saith, the Virtue thereof is not sufficient, but by long time to dissolve Metals.

XXIII.   (Raymundus saith in the end of his natural Magick, that there is a Salt made of the Lees or Tartar of Wine, or *Aquae Vitae,* called the Salt of Art and Mercury, without which Salt (saith he) there is nothing can be done: Also he begins his Practice with this Salt in the first and last Chapter of his Codicil.)

XXIV.   Wherefore as the same Philosopher affirms, among these things is this Menstruum one of the Secrets of this Art, whose Virtue must be increased by a wise management of the Matter: you must circulate this Menstruum in the Unctuous humidity in a Vessel of Circulation, by rotation continually, an hundred and twenty days, in the hottest Fornace.

CHAPTER LXX
THE REMAINING PROCESS OF THE VEGETABLE STONE.

I.   Hitherto the Process of the Vegetable Stone has been long and Obscure; but that nothing may be doubtful to the prejudice of my professed Love to your Lordship, I say that all these things spoken by Raymundus are covered with the Mantle of Philosophy. Truly his intention is, that there should be made a dissolution with the Spirit of Wine, but that this Spirit of Wine should be joined with another Menstruum resoluble, without which Resolution can never be attained.

II.   *(Here the two Spirits are joyned together, the Vegetable Menstruum or White Oyl of Tartar,[4] and our Metalline Oyl).*

III.   And that Menstruum Resoluble is generated only of a Metalline kind: for it is a potential or mighty Vapour, being in every Metalline Body, joyning together two extreams, *Sulphur* and Argent *Vive*.

IV.   And so indeed after this sort, Our water is a Metalline water, which because it does favour of the Nature of either extream, it therefore brings our

---

[4] *Is this alcohol? -HWN*

Resolutive Menstruum into Act.

V. Now how this Menstruum, which is Unctuous, Moist, Sulphurous, and Mercurial, agreeing with the Nature of Metals, and wherewith Bodies must be Artifically dissolved, may be had, we will here shew by clear practice.

VI. Take the sharpest humidity of *Grapes,* and in it being Distilled, dissolve the Body, well Calcin'd into a Redness, into a Chyrstalline clear and Ponderous water, the which Body Calcin'd into Redness, is of the Masters of this Science called *Sericon.*

VII. (Now comes in the Practice of *Pupilla,* of the dissolution of the *Red Lyon,* for the fire of nature, called also *Red Lead, Red Coral, Sericon* is of the Nature of Black Pepper, Euphorbium, etc. of a hot biting and fiery Nature, all which things are spoken only by way of Comparison.)

VIII. Then of this Crystalline water, let there be made a Gum, the which in Taste will be like to *Alum.* This Gum by Raymundus is called *Vitriol Azoth,* from which let there be drawn with a gentle Fire, first a weak water, with no more Taste or sharpness than simple Well water. *(Fresher water there is none in Taste, yet will it never Consume or Waste, tho' it be used never so often; nor will it be ever less in quantity).*

IX.   And when the White Fume shall begin to appear, change your Receiver, and Lute it strongly, that it breath not forth; so shall you have our burning water, Our *Aquae Vitae,* and Resolutive Menstruum, (the which before was Resoluble) a Vapour potential, a mighty Vapour, able to dissolve Bodies, to Putrifie, and to Purifie, to divide the Elements, and also to exalt the Earth into a wonderful Salt, by the force of its attractive Virtue. This is our Fire of Nature.

X.   This water has a bitter sharp Taste upon the Tongue, and also a kind of stinking Menstruum: and because it is a water which is very Spirituous and Volatile, therefore within a Month after it is Distilled, it ought to be put upon its Calx. When it is Affused upon the Calx, it will without any external Fire, boil if the Vessel be closely shut; and it will not cease to Ferment or work, till it be all dryed up into the Calx.

XI.   Therefore you must put no greater a quantity of it to the Calx, but what may just cover it as it were and so proceed, *(When the Fornace is dryed up)* to the whole Complement thereof, (as in the Operation of the Compound water,) and as the work requires.

XII.   And when the Elixir shall be brought into a Purple Color, then let it be dissolved with the aforesaid Vegetable Menstruum into a thin Oyl, the same Menstruum being first rectified, and let the same by the Circulation of the Spirit of our water be fixed; so will it have Power to transmute or change all Bodies into pure Gold, and to Heal and Cure all Infirmities and Diseases in Man's Body, ten thousand times better than all the Potions and Prescriptions of Galen or Hippocrates.

XIII.   This Elixir is the true AURUM POTABILE, and no other; for it is made of Gold Elementated and Circulated by the spirituous wheel of Philosophy; and it is so wrought with the Air, Gass, potency, or spirit of Mercury dissolved by its self, sublimed and rectified, as that the body of Gold by it may not only be curiously and exquisitely Elixirated; but also that it may then afterwards be brought to such a perfection by this our Art, as to be applied profitably to the most desirable work.

XIV.   Thus you may see, we have hid nothing concerning this our desired Elixir of the Vegetable Stone: I shall now proceed to that of the Animal Stone, which is but a work of three days; and in three days will be completely ended. My advice to you is, not to gather the Leaves of Words; but the

Fruits of Works, the profit of the things fought after.

XV. And know that in this Work, I have not so much affected the Curiosity of Language, or Elegancies of Stile, as the denudating the Essential Verity, and exposing the very Power of Truth to your View, which by reason of my haste I have now concisely done in few words.

CHAPTER LXXI
OF THE ANIMAL STONE.

I. We now come to reveal the most noble and High Secret of Secrets, viz. the Mystery of our Animal Stone desired of all Mankind, and the way and manner of its Elixiration. But this Animal Elixir is neither from Wine, as it is Wine, nor from Eggs, Hair or Blood, as they are such things, but only from the Elements: And these Elements we ought to search out, in the Excellency of their exceeding Simplicity and Rectification.

II. The Elements as *Roger Bachon* saith, are the Roots of all things, the Mothers of everything: yet the Elements of the said things do not enter into the Work of this Our Elixir; but only by the Virtue and Commixtion of those Elements, with the Elements of Spirits, and Bodies of Metals.

III. Yet so indeed as Roger Bachon saith, the Elements of those things aforesaid do so enter in as to pierce through (tho' not to dwell there) and to Accomplish this Our great Elixir.

IV. Notwithstanding among all those things which be Natural, (as all the rest be, which

Philosophers have taken,) there is one thing yet, which is found more pretious, more excellent, more proper, and more Natural than all the rest, for this our purpose.

V. The which one thing, because it is more excellent than all the rest, the Philosophers have taken for the nearest; because of the singular perfection which God has given to the *Microcosim* or lesser World, in whom are not only the Idea's of the Courses and effects of the Planets, Stars, and Asterisms, but also the Complexions, humours, Spirits, and Natural Virtues of the Elements.

VI. And therefore consider the most noble Bird of *Hermes,* which when the *Sun* is in *Aries*, begins to fly; and as it is advised, so let it be brought forth and sought for. Seek out the true *Sulphur* from his *Mine* or *Minera,* not being corrupted, for the whole perfection lies in the uncorrupt *Sulphur.*

VII. This is our Stone, the which as *Aristotle* saith, in his Secret of Secrets, is generated in the Dunghil, Highways, and must be divided into four parts: because saith he, each part has one Nature, the which parts must be joyned together again, till they resist or strive no more; when they are joyned unto it, it shall be White; if Fire, Red; as you please.

VIII. But understand, that this Division, must not be a Manual Division, (but in Power and Effect,) wherefore, let this one thing which all Men have (its over-flowing Flegmatick property being somewhat Evacuated) be put into Kemia or proper Vessels, which Seal up Philosophically; let it putrifie in a moist Fire a long Season, into a black thickness.

IX. Then by the second Degree of Fire, let it be Coagulated into a dryness, after many Bublings which it will make, wherein shall shine innumerable Colors: and when all that which is fine and subtil, shall Ascend upwards (or *Sublime*) in the Vessel most White, like as the Eyes of Fishes, the work is compleat in the first part.

X. This truly is a marvelous thing, more to be wondred at, than any Miracle of Nature, for then the self same White, has fully the Nature of White Sulphur, not Burning *(or Silver,)* and is the very *Sulphur* of Nature and *Argent Vive*.

XI. Let some quantity of Luna be added to it in the manner of an Amalgama; then it brings forth, by Operation, or generation of White into White; and the same thing worketh it into Red, and is made compleat into Red, by a greater Digestion in the Fire.

XII. Then, as the Philosophers advise, let the two Sulphurs, viz. the White and the Red be mingled with the Oyl of the White Elixir, that they may work the more strongly; upon which, if the Quintescence of the Vege-table Stone shall be fixed, you shall have the highest Medicine in the World, both to Heal and Cure Humane Bodies, and to transmute the Bodies of Metals in to the most pure and fine Gold and Silver.

CHAPTER LXXII
THE RESERVED SECRET EXPLICATED.

I. And now we are drawing near to the end of this work, we shall hereunto add and Explicate one Secret, even our reserved Secret, hitherto Buried in the Abyss of Aenigma's and deep Silence.

II. We say that the Body of the Volatile Spirit, fixed, by Fire against Nature, ought to be dissolved in the Vegetable Water, that is to say, in our Vaporous Menstruum; not in water of the Cloud, but in water of the Philosophers.

III. In which Dissolution, the Body is made light, for its more pure and subtil part is lifted up (*or Sublimed*) from Salt and Combustible Faeces, by Virtue of the water attractive: which is more clear than the water of the *Margarite,* as I have seen.

IV. And of this substance Fermented with the Oyl of *Luna* or *Sol,* is made the great Elixir, for the transmutation of imperfect Bodies.

V. It must oftentimes be dissolved and Coagulated with its Ferment, that it may work the better; and with this said Mercurial substance, thus Elevated (or sublimed) we Counterfeit the most pretious Margarites or Pearls, not inferior to the sight, to the very best that ever Nature produced.

VI. And with these Artifical pretious Stones, we shall finish the discourse of Our pretious Stones, (Mineral, Vegetable, and Animal) the abscondite Mysteries of which, being by the Wise and upright Sons of Art prudently kept Secret.

VII. Pray the most Good and Gracious God, to open and reveal the same, at one time or another, even as it shall please him, to his despised Servants and little ones.

VIII.   O most incomprehensible light, most Glorious in Majesty, who with the Charity of thy Heavenly Rays dost Darken our Dimmer Light; O Substantial Unity, the Divine three, the joy and Rejoycing of the Heavenly Host, the Glory of Our Redemption.

IX.   Thou most Merciful, the Purifier of Souls, and the perpetual subsistance; O most Gratious, through daily Dangers and Perils which thou suffers us to undergo, and through this Vexatious vail of Vanity, bring us to thy heavenly Kingdom.

X.   O Power and Wisdom, thou goodness inexplicable,

uphold us daily, and be Our Guide and Director, that we may never displease thee all the days of our Lives, but obey thee, as Faithful Processors of thy Holy Name.

XI.   Early, even betimes O Lord, hear thou my Prayers, by the Virtue of thy Grace, help forward my desires, and enable me I beseech thee to perform they Holy Will.

XII.   O most excellent Fountain, boundless in Treasures, thou scatterest thy good things without measure amongst the Sons of Men, and thou makest every other Creature to partake of thine especial kindness.

XIII.   Thou are worthy O Lord, to behold the Works of thy Hand and to defend what thy Right Hand has planted, that we may not live unprofitably, nor spend the course of our Years in Vanities.

XIV.   Grant therefore we beseech thee, that we may live without falshood and deceit, that avoiding the Great danger of a sinful course of Life, we may escape the Snares of Sin.

XV.   And as I Renounce the Loves of the things of this Life and the Concupiscences or Lusts thereof, so accept of me thy Servant, as a true and Spontaneous

Votary, who wholly depends on thy goodness, with all Confidence, possessing nothing more.

XVI. We submit ourselves to thee, for so it is fit; vouchsafe thy Light to discover to us the Immortal Treasures of Life; shew us thy hidden things, and be merciful and good unto us.

XVII. Among the rest of thy Servants who profess thy Name, I offer my self with all humble Submission; And I beseech thee O Lord, to forgive me, if I open and reveal thy Secrets to thy Faithful Servants. Amen.

CHAPTER LXXIII

RIPLEY'S PHILOSOPHICAL AXIOMS OUT OF THE THEATRUM CHYMICUM.

I.   Our Stone is called the *Microcosm;* One and Three; *Magnesia* and *Sulphur* and *Mercury,* all proportioned by Nature herself. Now understand that there are three MERCURIES, which being the Key of the whole Science, Raymundus calls his Menstruums, without which, nothing is to be done in this Art: but the Essential *Mercury* of the Bodies is the chief material of our Stone.

II.   Our Stone is a Soul and a substance, by which the Earth does receive its splendor: what other thing is *Sol* or *Luna* than a *Terra Munda,* a pure Earth, Red and White? The whole Composition we call *Our Plumbum* or *Lead,* the Quality of whose splendor proceeds from *Sol* and *Luna.*

III.   No impure Body, one excepted, which the Philosophers vulgarly call the *Green Lyon,* (which is the Medium which Conjoyns the Tinctures between *Sol* and *Luna* with perfection) does Enter into our Magistry.

IV. These Menstruums you ought to know, without which no true Calcination, or natural dissolution can possibly be done. But our principal Menstruum may be said indeed to be Invisible or Spiritual; yet by the help of our AQUA PHILOSOPHICA SECUNDA, through a separation of the Elements, in form of clear water, it is brought to light, and made to appear.

V. And by this Menstruum with great Labour is made the Sulphur of Nature, by Circulation in a pure Spirit; and with the same you may dissolve your Body after divers manners: and an Oyl may be extracted therefrom, of a Golden Color, like as from Our Red *Lead*.

VI. 1. DE CALCINATIONE: Calcination is the Purgation of our Stone, restoring it to its own Natural Color, inducing first a necessary dissolution thereof, but neither with *Corrosives,* nor fire alone, nor A. F. now with other Burning waters, or the Vapour of *Lead,* is our Stone Calcined; for by such Calcinations, Bodies are destroyed, for that they diminish their humidities.

VII. Whereas in our Calcination the Radical humidity is Augmented or multiplied, for like increases like, he which knows not this knows nothing in this Art. Joyn like with like, and kind with kind, as you ought;

every seed answers and rejoyces in seed of its own kind: and every Spirit is fixed with a Calx of its own kind or Nature.

VIII.   The Philosophers make an Unctuous Calx, both White and Red, of three Degrees, before it can be perfected, that shall melt as Wax, till which it is of no use. If your water shall be in a right or just proportion with your Earth, and in a fit Heat, your Matter will Germinate, the White together with the Red, which will endure in a perpetual Fire.

IX.   Make a Trinity of Unity, without dissention; this is the most certain and best proportion: and by how much the lesser part is the more spiritual, by so much the more easily will the dissolution be performed: drown not the Earth with too much water, lest you destroy the whole Work.

X. 2. DE DISSOLUTIONE: Seek not that in a thing which is not in it, as in Eggs, Blood, Wine, Vitriol, and the other middle Minerals; there is no profit to be had in things not Metallick: In Metals, from Metals, and by or through Metals, Metals are made perfect.

XI.   First make a Rotation of all the Elements; and before all things, convert the Earth into water by dissolution: Then Dissolve that Water into Air, and then make that Air into Fire: this done, reduce it again into Earth, for otherwise you labour in vain.

XII.   Here is nothing besides the Sister and the Brother; that is, the Agent and the *Patient, Sulphur* and *Mercury*, which are generated Co-essential substances. The dissolution of one part of the Corporeal Substance, causeth a Congelation of another part of the spiritual.

XIII.   Every Metal was once a Mineral Water, wherefore they may all be dissolved into Water again; in which Water are the four repugnant Qualities with diversity. In one Glass all things ought to be done, made in the form of an Egg, and well closed.

XIV.   Let not your Glass be hotter than you can endure your naked Hand upon, so long as your matter is in dissolution: When the Body is altered from its first form, it immediately puts on a new form.

XV.   3. DE DISPOSITIONE: Beware that you open not your Glass, nor ever move it, from the beginning of the work to the end thereof; for then you will never bring your work to perfection. Dry the Earth till it becomes thirsty in Calcination, otherwise you Act in vain. Divide the matter into two parts, that you may separate the subtil from the gross, or thin from the thick, till the Earth remains in the bottom of a Livid Color.

XVI. One part is Spiritual and Volatile; but they ought all to be converted to one matter or substance. And distill the Water, with which you would Vivifie the Stone. till it be pure and thin as water, shinning with a Blew Livid Colour, retaining its Figure and Ponderosity: with this Water *Hermes* moistens or waters his Tree, whilst in his Glass, and makes the Flowers to increase on high.

XVII. First divide that, which Nature first tyed together, converting the Essential Mercury into Air, or a Vapour, without which natural and subtil separation, no future Generation can be compleated.

XVIII. Your Water ought to be seven times sublimed, otherwise there can never be any natural Dissolution made; nor shall you see any Putrefaction like Liquid Pitch; nor will the Colors appear, because of the defect of the Fire Operating in your Glass.

XIX. 4. DE IGNIBUS: There are four kinds of Fires which you ought to know? the Natural, the Innatural, that contrary to Nature, and the Elemental, which burns Wood: These are the fires we use, and no others.

XX. The Fire of Nature is in everything, and is the third Menstruum. The Innatural Fire is occasionally so called, and it is the Fire of Ashes, of Sand, and Baths for putrefying: and without this no Putrefaction can be done.

XXI. The Fire against Nature, is that which tears Bodies to pieces or Atoms; which is the fiery Dragon, violently burning like the fire of Hell. Make therefore that your fire within, in your Glass, which will burn the Bodies much more powerfully than the vulgar Elemental fire can do,

XXII.   5. DE CONJUNCTIONE: Conjunction is the joyning together of things separated, and of differing Qualities; or the Adequation or bringing to an equality of principles: he which knows not how to separate the Elements, and to divide them, and then to conjoyn them again, errs, not knowing the true way.

XXIII. Divide the Soul from the Body, and get that, for it is the Soul which causes the perpetual Conjunction: the Male, which is our Sol, requires three parts; and the Female which is his Sister, nine parts; then like rejoyces with like forever.

XXIV. Certainly Dissolution and Conjunction, are two strong principles of this Science, tho' there may be many other principles besides.

XXV.   6. DE PUTREFACTIONE: The Destruction of the Bodies is such, that you are diligently to Conserve them in a Bath, or our Horse-Dung, viz. in a moist heat for ninty days Natural: but the Putrefaction is not completely Absolved, and brought to whiteness, like the Eyes of Fishes, in less than 150 days; the blackness first appearing, is the Index or Sign, that the matter draws on to Putrefaction.

XXVI.   Being together Black like Liquid Pitch, in the same time, they swell and cause an Ebulition, with Colors like those of the Rainbow, of a most beautiful aspect; and then the water begins to whiten the whole Mass.

XXVII.   A temperate heat working in moist Bodies, brings forth blackness, which having obtained, there is nothing that you need fear, for in the same way, the Germination of our Stone does follow, and forthwith, to wit, in the space of thirty *(or Forty)* days, you have *Gas,* or *Adrop,* which is our *Uzifer* or *Cinnabar*, and our *Red Lead*.

XXVIII.   Take heed to defend your Glass from a Violent Heat, and a sudden Cold; make use of a moderate Fire, and beware of Vitrification.   Beware how you bind up your matter; mix it not with Salts, Sulphurs, nor the middle Minerals; let Sophisters prate what they will, Our Sulphur and our Mercury are found in Metals only.

XXIX. 7. DE COAGULATIONE: Coagulation or Congelation is the induration or hardning of things, in *Calore Candido,* and the fixing of the Volatile Spirit. The Elements are forthwith converted, but the Congelation is no way impeded, for those things which are Congealed in the Air, melt or soften not in the Water; for if so, Our work would be destroyed, and come to nothing.

XXX. When the Compositum is brought to Whiteness, then the Spirit is United and Congealed with the Body; but it will be a good length of time before such a Congelation will appear in the likeness or Beauty of Pearls. The cause of all these things is the most temperate heat, continually working and moving the Matter. Believe me also, that your whole Labour is lost, except you revivifie your Earth with the Water, without that you shall never see a true Congelation.

XXXI. This Water is a Secret drawn from the Life of all things existing in Nature; for from Water all things in the World have their first beginning, as you may easily perceive in many things. The substance or Matter is nourished with its proper Menstruum, which the Water and the Earth only produce, whose proper Color is Greenness.

XXXII.   Understand also that our fiery Water thus actuated is called the Menstrual Water, in which Our Earth is dissolved, and naturally Calcined by Congelation; when you have made seven Imbibitions, then by a Circumvolution, putrifie again all the Matter without addition, beholding in the first place the blackness, then the Whiteness of the Congealed Matters.

XXXIII.   Thus your Water is divided into two parts: with the first part, the Bodies are purified: the second part is reserved for Imbibitions; with which afterwards the Matter is made black, and presently after with a gentle fire, made White, then reduce to Redness.

XXXIV.   8. DE CIBATIONE: Cibation, is the Feeding or Nourishing of our dry Matter with Milk and Meat, being both administred moderately, till it is reduced to the third Order: you must never give so much as to cause a suffocation, or that the Aqueous humour should exceed the Blood: if it drinks too much, the work will be hurt.

XXXV.   Three times must you turn about the

Philosophick Wheel, observing the Rule of the said Cibation, and then in a little time it will feel the Fire, so as to melt presently like Wax.

XXXVI.   9. DE SUBLIMATIONE: Sublime not the matter to the top of the Vessel, for without Violence, you cannot bring it down to the bottom again; by a temperate heat below, in the space of 40 days, it will become black and obscure. When the Bodies are purified, let them be sublimed by degrees more and more, till they shall be all elevated or converted into Water.

XXXVII. We use Sublimation for three Causes. First, that the Body may be made spiritual. Secondly, that the Spirit may be made Corporeal and fixed with it, and become Consubstantial with it. Thirdly, that it may be purified from its Original Impurities; and its Sulphurous Salt may be diminished, with which it is infected; subliming it to the Top, as White as Snow.

XXXVIII.   10 DE FERMENTATIONE: Fermentations are made after divers manners, by which our Medicine is perpectuated. Some dissolve *Sol* and *Luna* into a certain clear Water; and with the Medicine of them, they make the same to Coagulate, or be Coagulated, but such a Fermentation we propose not.

XXXIX. This only is our Intention, that first you must Break, or Tear, or Grind the matter to *Atoms,* before you Ferment it: Mix then presently your Water and Earth together; and when the Medicine shall flow like Wax, then see the above mentioned Amalgamation and put forth the same; and when all that is mixed together, above or on the top of the Glass, (being well closed,) make a Fire, till the whole be Fluxed; then make projection as you shall think fit, because it is a Medicine wholly perfect: Thus have you made the Ferment both for the Red and the White.

XL. The true Fermentation is the Incorporation of the Soul with the Body, restoring to the same the Natural Odour, Consistency, and Colour, by a Natural Inspissation of the separated things. And as the Magnet draws Iron to itself, so our Earth by Nature draws down its Soul to itself, Elevated with Wind: For without doubt, the Earth is the Ferment of the Water, and by Course or Turns, the Water is the Ferment of the Earth.

XLI. We make the Water most Odoriferous, with which we reduce all the Bodies into Oyl, with which Oyl we make our Medicine flow. We call this Water a Quintessence, or the Powers, and it Heals or Cures all humane Diseases. Make therefore this Oyl of Sol and Luna, which is a Ferment most fragrant in smell.

XLII. 11. DE EXALTATIONE. Exaltation differs a little from Sublimation, if you understand aright the words of the Philosophers. If therefore you would Exalt your Bodies, sublime them first with Spiritus Vitae; then let the Earth by subtiliated by a Natural rectification of all the Elements; so shall it be more pretious than Gold, because of the Quintessence or Powers which they contain.

XLIII. When the Cold does overcome the Heat, then the Air is converted into water, and so two contraries are made by the way, till they kindly conjoyn and rest together: after this manner you must work them, that they may be Circulated, that they (one with another) may speedily be Exalted together in one Glass well Sealed, all this Operation is to be done, and not with hands.

XLIV. Convert the Water into Earth, which will quickly be the Nest of the other Elements; for the Earth is in the Fire, which rests in the Air. Begin this Circulation in the West, then continue it till past the Meridan, so will they be exalted.

XLV. 12. DE MULTIPLICATIONE. Multiplication is the thing which makes the augmentation of the Medicine in Color, Smell, Vertue, and Quantity; for it is a Fire, which being Excited, never dies, but always dwells with you, one spark of which is able to make more Fire by the Virtue of Multiplication.

XLVI. He is rich which has but one Particle or Grain of this our Elixir, because that Grain is possible to be augmented (by one way) to Infinity: If you dissolve this our dry Pouder, and make a frequent Coagulation thereof, you will augment it, and so you may Multiply it, till it increases in your Glass, into the form of a Tree, and which Hermes calls a Tree, most Beautiful in Aspect. Of which one Grain may be Multiplied to an hundred, if you know how wisely to make your Projection.

XLVII. Our Elixir, the more fine and subtle it is made, so much the more compleatly it tinges, and disperses its Tincture. Let your Fire be kept equally close, Evening and Morning; so much the longer you keep the Fire, so much the more profitable it will be; and Multiply more and more in your Glass, nourishing your Mercury in its enclosure, whereby you will have a greater Treasure than you could desire.

XLVIII. 13. DE PROJECTIONE. If your Tincture be true and, not Variable, you may prove it in a small quantity thereof, either in Metal or Mercury: It cleaves thereto as Pitch, and so Tinges in Projection, that it is able to endure the strongest Fire: But many through Ignorance destroy their work, by making Projection upon an impure Metal.

XLIX. See that you Project your Medicine upon your Ferment, so will that Ferment be Brittle as Glass: Project that Brittle Medicine upon pure Bodies, so have you Silver or Gold, enduring the severest Test.

L. Give not liberty to the Reins lest you sin, but Religiously Fear and serve the Lord your God; think yourself always before the Tribunal of the most high, the great Judge and Rewarder of Mankind, who will return to every Man according to his works.

LI. 14. RECAPITULATIO. Take heed diligently to the Latitude of our Stone, and begin in the Occident, where the Sun sets, where *the Red Man and White Wife* are made one, conjoyned and Married by the Spirit of Life, that they may live in Love and Quietness.

LII. The Earth and Water, are joyned in a fit proportion; one part of Earth or Body to three of Spirit, which is 4 to 12. and is a good proportion: you must take three parts of the Female to one of the Male: by how much less there shall be of the Spirit in this Dispensation, Conjunction, or Marriage, by so much the sooner will the Calcination be Absolved.

LIII. The Calcination performed, then you must dissolve the Bodies, divide, and Putrefie them; and all the Secrets of our other lower Stars will have a perfect Coherence and understanding with the Poles of our Heaven, and will appear with inexplicable Colors of Light and Glory, Transcending in Lusture and Beauty, all other things in the World, and all this before the perfect Whiteness.

LIV. And after the perfect Whiteness, you will have a Yellow, the false Citrion Colour: afterwards the Blood Red, unchangable forever, will be manifest; so have you a Medicine of the third Order in its kind, which may continually be Multiplied. But this you must not in the least be Ignorant of, that the RED MAN does not Tinge, nor yet his WHITE WIFE, till they themselves are first Tinged with our Tincture or Stone.

LV. When therefore you prepare your Matter by this our Art; hide your Bodies all over, and lay open their Profundities or Insides, destroy the first quality of all your Materials, and the more excellent second qualities, which in these you must separate; and in one Glass, and by one Government and Order, convert the four Natures into one.

LVI. The Red Elixir must be divided into two parts,

be-fore it be Rubified, which put into two Glasses; and if you would have a double Elixir, one of Sol and another of Luna, do thus:

LVII. With Mercury multiply presently the Medicine into a great quantity, if you have at first only so small a quantity as a Spoonful: then may you multiply them together into a White and Red Medicine, which by Circulation you must convert into a perfect Oyl according to our directions; and this Multiplication from your first small quantity may be continued, should you live a thousand Years. These Oyls will fix Crude Mercury into perfect Sol and Luna.

LVIII. This pure and fixed Oleaginous substance, Raymundus calls his *Basillisk,* whose Explication is so easie to be understood, that it needs no more Words.

LIX. For our Metals are nothing else, than our two *Minerae,* viz. those of *Sol* and *Luna,* as Raymundus wisely Notes; The Splendor of *Luna,* and the Light of shining *Sol*. In these two *Minerae,* the Secret dwells; tho' the Splendor may for a while be hid from your Eyes, which by the help of Art, you may easily bring to light.

XL. This hidden Stone, this one thing, purifie it,

wash it in its own Liquor, Water or Blood, till it grows White; then prudently Ferment it, so have you the Summ and Perfection of the whole Work.

- **FINIS** -

LIBER
Secretisimuss

BY: GEORGE RIPLEY

PRODUCED BY:

RAMS

1982

LIBER SECRETISSIMUS

GEORGE RIPLEY

Aol hic aut nusquam.

Liber Librum aperit.

The Whole Work of the Composition of the Philosophical Stone and Grand Elixir, and of the First Solution of the Grosse Bodies

Take our Artificial Antimony, but not the Natural Antimony as it comes out of the Earth, for that is too dry for our work, and hath little or no humidity, or fatness in it, but take I say, our Artificial Antimonial Compound, which is abundantly replinished with the Dew Of Heaven and the fatness and unctuosity of the earth, wherein precious Oils and rich Mercuries are by Nature closely sealed up, and hidden from the eyes of all ignorant deriders of the great and wonderful mysteries of Almighty God, to the end that seeing they should not see, nor understand, what he hath in-closed in the most obvious, common, and contemptible beginnings of all Things in the whole World.

This our Antimonial Compound is only to be revealed to the Children of Art, who firmly believe the constant truth thereof, and whom in all fraternal love and charity we say, that it is made of one Sulphur, and of two Mercuries, which otherwise by the wise Philosophers are called, the Sun, Moon, and

Mercury, or as some of them will more plainly have it, Salt, Sulphur, and Mercury, which are the three several and distinct substances and bodies, although for the most part we term them but one Thing, because in the conclusion of our work they make but one Thing, that is our admirable Elixir, and they have all one original, and tend altogether but to one end. For if we had not in our Work a triune aspect of these Planets, and did not begin it with a Trinity, all would be lost labour and inutilous profile.

Wherefore if thou wilt thrive in our Art, we wish thee to begin with our Mineral Trinity, whereof this our Artificial Antimonial compound is made. Take then first in the prime beginning of thy Work, these three noble Kinsmen, who are immediately indued with all the strong and subtile qualities of the four Elements, and in their due and most natural proportions, (in which proportions see thou do not erre, for if thou do, thou shalt never reduce those bodies into our true Chaos, and so thou wilt be constrained to begin again, which will be a most tedious discouragement unto thee). Put them into a good and strong cucurbit, or glass body, and close it well on the Top, that none of the spirits exhale, for if they find a Vent to evaporate, thou art undone, because thereby thou loosest and wasteth the flowers of our Gold. When thy Vessel is well closed, put it into the Philosophers Oven, and set it in Ashes or sand, with a temperate fire under it, for the space of a Philosophers Month, which is six whole weeks, and in that time our grosse bodies will be dissolved and mortified and made fit to begin a more royal generation.

In this time of dissolution and putrefaction our three noble Kinsmen, most unnaturally become the immane homicides of each other, for they spare not with all cruelty to extract each others vital blood, and are stewed in their own proper gores, and become soft and tender, like unto butter, and are made all one thing without any difference, or distinction. When thou hast brought thy work to this pass, thank God, and be glad that through his Grace and mercy thou hast obtained our Chaos dark and mistie, which is the true one Thing written of by all the Philosophers, our confused mass, and the prime ground of all our Secrets, for therein lieth invisible couched, our Gold and Silver, our Sulphur and Mercuries, our Christalline Water, our Oils and Tinctures, and our four Elements which thou must make visible and apparant to all seeing eyes, else can nothing be effected, neither shalt thou ever obtain thy wished for silvery and golden desire.

These Mercuries, Waters, Oils, Tinctures, and Elements, make visible then and conspicuous thus. After the aforesaid months end, thy vessel being cold, open the mouth thereof, and set on the top thereof a head of glass well fastened thereunto, and place it in our Bath, and close well a receiver to the mouth of the helment, and draw out all the insipid and faint water, which take away and reserve it close by itself, then fix your receiver on again, and with a stronger fire in ashes, and draw out all the White fume, which is called our Air, silverie Tincture and Virgins Milk, which also remove and keep it likewise most close stopped by itself. Then last of all put to another receiver, and in sand, with the strongest fire thou canst make, separate the red fume, which is called our natural fire, our golden Tincture, and radical humidity of our Elemental bodies, and continue thy fire so long until it leave bleeding, then asswage the fire by degrees, and suddenly close it well with wax, that the spirits vanish not away, for this is called our blessed Liquor, and trust me there is not a stronger poison in all the World than it, therefore keep it close and meddle not with it till hereafter.

Thus now the work of Art, for Division and Separation, is the sole work of Art and of the Artist, and not of Nature; for here Nature is forced by the Skill of the Workman, to forgo and part with her beloved Elements, which she so straightly kept chained and inclosed in her bosome, and which by violence by external fire, are even as it were forcibly rent and torn from her.

Indeed, the first Work of Solution and Mortification is the Sole operation of Nature, for the Materials being inclosed in their dungeon there Nature attenuates them, there she dissolves and putrefies them, and causeth them to run a retrograde course into their own first matter again, that is, into a slimie and viscous mass and substance, whereof at first out of the four Elements they were formed and begotten. And in that first work, the Artist is but only a fire maker, which only giveth Nature strength and power to work in and upon the materials, for without an exciting fire Nature were not able to effect anything, but she would wholly remain idle and void of course, and that in regard of the extreme coldness of the Mercury, for where cold predominates, heat must needs lie fettered and immoveable, but when the external fire stirreth up and provoketh the fierce indignation of the Salt Menstrue, she presently seizeth upon the fat unctuous body of the Sulphur, and extracts his natural heat and fiery element, and then both of them together do so ardently incircle and work upon the cold Mercury, that they dissever and scatter all his members into smaller atomes than fly in the beams of the Sun, and then heat beginneth to get the supereminence.

    Therefore in this FIRST WORK the Operator is but the Trustie administrator of Nature, which in this case, the most illeterate and simple man in the World may be.

    But in the Secret Work of the conjunction of our Elements, both Nature and Art, hand in hand accompanying each other, for there the Artist findeth and imbibeth, and Nature fixeth and congealeth, which we will show hereafter, when we handle that work.

The FIRST WORK OF SOLUTION reduceth our Trinity and Antimonial Compound into a green gum, called the Green Lyon, which gum dry moderately well, but beware thou burn not his flowers, nor destroy greenness, for therein lieth his Soul, which is our main Secret.

And our SECOND WORK manifests those things which were before hid from sight, namely, it makes our four Elements to appear visibly, and to be had generally and distinctly. But in this Second Work if thou extract our Air and our Fire with the phlegm water, they will the more naturally and easily be drawn out of their infernal prison, and with less losse of their Spirits, than by the former way before described.

After division is perfectly celebrated, thou shalt find in the sides of the Cucurbit, and also in the head of the Alembick a white hard Ryrne much like the Congelation of a frosty vapour, or like unto Mercury sublimate, which keep charily by itself in a clean glass well closed, for therein lieth hid a great secret, for therewith thou mayest abbreviate the work, in half the time, and with half the labour that else it would be done in, which will prove a greater ease and benefit, and shorten expenses.

The Cleansing of our Base.

It is most certain, that the external fire is a great friend to us, and his nature is such, that he can indure no impurity in anything, therefore at the day of Judgement, the Elemental fire shall purifie, cleanse, and burn up all the impure earth that we tread on, and purge it from all dross and filth.

So likewise must it do in our work. Wherefore after the separation of our Water, Air and Fire out of our Earth; Take out the black feces which remain in the bottom of our Vessel, called our Dragon, and grind it into small powder, and calcine it in two crucibles luted together, in a potters furnace, or in a glass, or wind furnace, until it look whitish or something grey, which Caix keep by itself., for it is called the Base and Foundation of our Work, and it is now called MARS, and our White Fixed Earth.

The Purification of our Faint Waters.

But that thou mayest loose no time nor be idle, while the Earth is calcining, distil the Water that was first drawn out of the Compound, seven times, until it be clear as Crystal, but do it by itself alone, not intermixing any other thing therewith, and then hast thou our pure river of Water of Life, which hath power and virtue to revive the dead body from whence it sprung, and to make it white and Shining like a pure Alabaster. As for our other Arie and fiery Waters, they are so fixed and permanent, that no fire will again elevate their substances, but they would stand in the fire, until Doomsday without any wasting or exhalation.

All things being now purified without and within, now beginneth the Work of Art and Nature, wherein, the one must inseparably aid the other, for if either refuses to help each other, all the former sweats and Toiles are to no purpose, for Nature can not bring our disserved elements together without the assistance of the expert Artist; neither can the Artist coagulate the elements put together without Nature, wherefore Nature implores the aid of the Artist with an Adjuna me, & ego adjunabose. And the Artist finding before, what good Nature had done him, in dissolving and putrefying the bodies, is as ready to help her, thus.

The Work of Conjunction of our Elements.

Take the abovesaid calcined faeces called MARS, or Our Dragon which devoureth his own Tail, and put thereof so much into a glass body, as scarcely fill half of it, and pour thereon so much of our rectified water of Life, as well may but cover the Calx made into powder, which done, incontinently stop the glass with a blind head well luted to the body, and set it in hot ashes, until the Calx have drunk up and congealed all the Liquor, which it will do in eight days, and so from eight days to eight days imbibe the said Caix with his own water, and that he will drink no more, but is very white and shineth as the Eyes of Fishes, and be full of Chrystalline Plates, then, the said Calx being very dry, take it out of the glass, and separate from it, all the Chrystalline Plates, and make them only into Powder, put that powder, which we call SULPHUR OF NATURE, or FOLIATED EARTH into another glass, and pour thereon some of our White Arie fume called Virgins Milk, upon a lent fire congeal them together, and continue this Work until it have drunk almost all his Air, and is become fixt and permanent. Then take it out, make it into powder and incerate it with part of his reserved Air by drops, until it become like liquid honey, and that it will melt and flow like Wax, on a coal fire, and not evaporate, thus hast thou the perfect White Stone, and Silverie Medicine, which transmuteth all imperfect metalline bodies into true, solid and perfect Luna.

The Red Work.

When thou hast once obtained our White Stone in manner and form aforesaid, divide it into two equal parts, and to the one of them put the 4th. part of Mercury sublimate both of them pulverized, mix them well together, and put them into a glass which stop close, and set it into your furnace, and there let it stand for the space of a month, in a temperate fire, until it be one body, then take part of it to project for your necessitie, and the other part you may still multiply with sublimate, or Quicksilver purified with Salt and Vinegar, unto your lifes end, and so substract and multiply at your pleasure.

But for the Red, take the other half of your reserved White Stone, and pulverize it, and put it into a glass, and pour thereon a little of our fiery water, or Golden Tincture, and congeal them together upon a slender fire lest your glass break, by force of the venome and insuperable power of our Red and fiery Mercury, do so once or twice, until it be perfectly fixed, then take it out make it into red powder, and incerate it in a crucible with his said red Oil, or fiery Water, until it flow like Wax, as you did the White Medicine, then have you OUR DARK RED STONE somewhat like the powder of an Hoematite, which is able to do miracles upon earth, but we intend not to reveal them at this time, leaving it to be experienced by those, whom Almighty God shall think worthy to teach, by this our little and brief, but pithie and true book, unto whom we perpetually owe all thinks and praise, for endowing us with the Knowledge thereof.

The Accurtation of the Great Work, which saveth half the Time and Labour.

The White Rogue, or powder, whereof we spake before, and willed thee carefully to reserve it, which is perfect Sulphur of Nature, and foliated Earth, which needs neither imbibing, nor digesting unto the white.

Take it therefore, and being ground fine and small, imbibe it with a fourth part of our before said Air, or Virgins Milk.

But observe by the Way, that thou must have great store of our Water, Air and Fire, and those extracted out of five or six several Compounds, or Chaos, so that after you have driven one Chaos out of the Oven, you must presently set in a new one, and so successively one after another, and then separate their elements, for else you will want waters and oils for imbibition, inceration, and multiplication, and if thy work be discontinued for want of such materials, all is spoiled and will come to nothing, for if thou once begin, thou must proceed without stay or interruption unto the full end.

But to the matter, having imbibed the said Ryme, congeal the whole on a soft fire, until it be drunk up, then imbibe and congeal it twice more until it be fixed, after that powder it, and incerate it, with some of our Air by drops, as thou didst thy white medicine before, until it flow like wax upon a red hot iron, and fly not away, thus shalt thou have the White Stone perfectly made in half the time, and with half the labour which is a precious Jewel, and a great Secret.

The Accurtation of the Red Work.

Take the whole, or the half, of this our White Stone, made of the said **Ryme**, and being pulverized, put it into a strong egg glass, and imbibe it with a little of our Red fiery Mercury, and set it on a weak fire for fear of breaking the glass, congeal it into a dry powder, then imbibe it and congeal it so twice more until it be strongly fixed, then take it out, pulverize it, and incerate it with our said fiery Oil by drops in a strong crucible, on a gentle fire until it flow like wax as is before said. Then hast thou the Red Stone perfect with less labour, expence of time and costs, for the whichever thank God.

This Secret was never before discovered by any of the Ancient Philosophers, for they were ever envious of their rare Mysteries, which we have now fully disclosed, for the honour of God, and for thy good, that thereby thou mayst perform holy Works of Charity and Mercy, plentifully supplying and relieving the fatherless and widdowers, redeeming prisoners and captives, especially such as suffer for our Blessed Lord and Saviour, Christ Jesus sake.

Our White Stone is Multiplied by reiterate imbibition, congealation, and inceration, with our Airie Virgins Milk, for the more and oftner you put that to it, the more it increaseth in quantity, and it is thereby made the more subtile and penetrating, and converteth the more metal, with the lesser of its quantity.

In like manner our Red Stone is also multiplied by reiterate imbibition, congealation, and Inceration with our fiery Oil, or Red Mercury, and therewith thou mayst so acuate it, that it shall be able, not only to penetrate metals, but also the hardest Stones, and whatsoever other said Things in the Whole World.

Whosoever then shall obtain these Medicines, he shall have incomparable Treasures, above all the Treasures of this World.

FINIS.

Of this Salt, Helbigius Saith:

British Museum MSS. Sloane #630

The above is the only title given this small tract

The Philosophers speak of MERCURY, SULPHUR and SALT, this third principle viz, the Salt while I breathe I will extoll with most ample prayers as most true principle. In this is whatever the Wise Men seek. Its all in all. It opens and shuts. This reigns in the Air. It rules both earth and water, which equal Sulphur and utmost force. Its the third and last of most natural things in Nature. Water is the Mother of it; This is the MERCURY sought after by Chemists, and their Sulphur; its neither Kitchen Salt, nor Sea Salt nor calcined Salt, nor volatilized Salt. Nor any other thing produced by the help of fire. But the most noble grain of the maturity and excellency of everything. Added out of the subtility of the superior Waters, to the Inferiour, for a soul and for the beginning added to them. We name it in einminent manner the Salt of Nature.

This primogenius Salt is not corrosive; not acrid, scarce sensibly astringent, most penetrative and opening, Dissolving and when you follow the Natural Process or motion, it is Coagulating and Maturating. Its part of the certain body obtained out of the fire, and through continued motion, so nobilitated, and maturated that it deservedly merits the Title of the Soul Essence and Salt of Nature.

It rests upon the whole Universe, but in one place in greater quantity than another, one part whereof is easier to obtain than another.

Whether is the MERCURIAL Viscous liquor or the SULPHUREOUS Salt easier obtained?

There are three Kingdomes: The Inferiour that lies under mens feet, the Middle Kingdom, the Animal, and the Superiour Kingdom, that, which is above us and in which we live; It is in all three as vide the Author.

But the middle Kingdom possesseth the most excellent SALT of Nature and the greatest part thereof & etc., which is as well drawn thereout as out of the Superiour and Inferiour Kingdoms.

The Seat hereof therefore is the Middle Kindom, whose seat I call SPELUNCA, a cave and Den, the house and habitation of sadness and gladness. The Inhabitant whereof or Indweller is called the Magnet, Chaos.

Improperly Sendivogius stiles, the Hyle, the first matter. And it enjoys so great (an) abundance of the Salt of Nature, as no body in the world hath so much. It is beheld by many, but by reason of its viscid slimy cloathing, and the darkness of its leaden colour, it seems vile, and being dug up is sometimes cast back again (rejected) with nausia and loathing. It is purchased with labour but never with money. And being once obtained it is always sufficient (the reason thereof is that it is so easily augmented).

The first begotten Salt of Nature being drawn down to the custody of the Magnet, and being purified and applied to another subject, by the heat of motion is stirred up and is made as it were an Agent, which dissolves the Metals and Minerals which are transformed by the motion of the Water, and it opens, enters, the more ignoble, crude, gross part of the transmuted Water; and it doth open that which incloses the Salt which is like itself and doth enter into it, and helps it then that perfect grave of essence may have greater power over the rest. E. Q. Gold.

Gold being dissolved radically (through the operation of this Salt as a Menstruurn) by the virtue of which, and also of his own essence (which is the same salt) which virtue is stirred up by the motion of external heat, thus the Gold is en-nobled, and is exalted to such a degree far exceeding the common and crude maturation it receives in the earth, so that it yields seed which inserted into the lesser and more unripe metals, by that Illumination doth advance them far above the excellency of common Gold.

This Salt is that Universal Menstruum, the Coagulated Water of Nature. The Vitriol of the Little World (that is the Philosophical Egg). The

Philosophical sublimate Mercury & etc. This is the field to which Gold serves for seed.

For the separation of this Treasure, First the unprofitable part must be cast away from the useful part, the useful part must be purified, and separated into two parts (viz. into Mercurial Water and Sulphurical Earth). The greater part flys away and ascends, the lesser part remains below as dead, until the rest of the filth being abstracted, the upper part doth exalt the lower; both which being united they ripen the Gold that is added to them, and effect it with infinite fertility.

This agrees with the other particular process which appoints to take crude Solar and lunar or Mercurial Minerals, wherein Mercury is as yet crude and unripe, wherein the sulphurial heat hath not yet so far coagulated the Mercurial Water; as to bring it to the body of Mercury, much less to the dryness of Antimony; (so that certainly its in neither of these, they are past it) purifie the matter by washing and removing the stony or too gross part, then putrify it 90 days, and then distil it in Balneo gently. The which all is rising with the heat, then you may see what more of the Mercurial Liquor rose, you may by

---

[5] Note. The Flowers that lie in the middle in the sublimation that are the reddest and shining are the best, not the lightest they are not fixed enough, and the lowest are too dropsycal, but experience will show this.

[6] Note. (or SULPHUR) some say by digestion of them together a due time.

greater heat of Sand come the Sulphureous Earth to arise in flowers.[5]

    Or the Flowers[6] may be got of another Solar Root.
Then by often Cohobation of this Mercurial Water upon this Sulphurial earth, viz, the flowers; the liquor is so impregnated with Solar SULPHUR, that it is the true Menstruum for pure Sol which will dissolve it and ripen it as above said, viz. Two parts of this Mercurial SULPHURATED liquor, and one part of purified Sol, or Solution of Sol, put into a Philosophical Egg, not a third full, hermetically sealed; placed in a nest in an Athanor in such a gentle heat as only keeps a continual vapour and moisture in the glass with a constant kind of motion, therein also; through all the colours, till it come to white and red & etc. Too much heat will too soon dry up the vapour and moisture, and fix the motion (and burn Icarus his wings) too little heat will not raise such natural Dewy Vapours, nor cause such Vegetable and Animal motions, and as of the outward, the like may be said of the Inward heat and moisture; the Philosophical Sulphur and the true Mercury of the Philosophers are the Inward fire and Water of the Philosophers, which for the same reasons must be exactly ordered in the due quantity aforesaid; And then the Red and White, one or both, as you please, may be again dissolved in the aforesaid

Menstruum, and coagulated till they attain the desired strength.

FINIS.

A Short Process Following This Treatise

The way to change MERCURY as red as blood which being converted into powder transmuteth LUNA into SOL; first make an AQUA-FORTIS of VITRIOL and NITRE and distill it 3 times from its own CAPUT MORTUUM.

Rx. of this AQUA-FORTIS ℥ iij g MERCURY V iij in a retort, let them distill then & cohobate so often until the MERCURY bears red as blood, which will be in 5 or 6 times, then bring it to a red powder by imbibing it with the red oil of VITRIOL 3 times, then dry it and bring it into a red powder and divide it into 8 parts.

Then take 1/2 ℥ SATURN upon a test, when it driveth put ℥ i g LUNA & 1/4 ℥ of SOL & i g the 8 parts of powder, and there will remain 1/2 ℥ of SOL.

A Word from the Publisher

Thank you for purchasing this small work from The R.A.M.S. Library of Alchemy. During his lifetime, Hans Nintzel was dedicated to the identification, acquisition, study, retyping and, when necessary, translation of what he considered to be the most important known works on Alchemy. Hans was assisted by his sparse network of fellow Alchemists, all members of the Restorers of Alchemical Manuscripts Society (R.A.M.S.). I was an active member of R.A.M.S.

My goal is to publish all of the works originally made available through R.A.M.S. as photocopies. To facilitate this, I have chosen to have the books professionally printed. I also have a few titles that I intend to add to the original R.A.M.S. Library, selected by strict criteria established by Hans.

If you have a work on Alchemy that you believe should be a part of the R.A.M.S. Library, please contact me through R.A.M.S. Publishing Company.

Philip N. Wheeler

CPSIA information can be obtained
at www.ICGtesting.com
Printed in the USA
LVHW021959100323
741362LV00007B/506